Henry Clark Houghton

Lectures on clinical otology, delivered before the senior class in the New-York homoeopathic medical college

Henry Clark Houghton

Lectures on clinical otology, delivered before the senior class in the New-York homoeopathic medical college

ISBN/EAN: 9783337274696

Printed in Europe, USA, Canada, Australia, Japan

Cover: Foto ©berggeist007 / pixelio.de

More available books at **www.hansebooks.com**

LECTURES

ON

CLINICAL OTOLOGY

DELIVERED BEFORE THE

SENIOR CLASS IN THE NEW-YORK HOMŒOPATHIC MEDICAL COLLEGE,

TO WHICH ARE ADDED

CASES FROM PRACTICE, AND SUMMARIES OF REMEDIES.

BY

HENRY C. HOUGHTON, M.D.,

SENIOR AURAL SURGEON TO THE NEW-YORK OPHTHALMIC HOSPITAL, PROFESSOR OF OTOLOGY IN THE COLLEGE OF THE NEW-YORK OPHTHALMIC HOSPITAL, PROFESSOR OF CLINICAL OTOLOGY IN THE NEW-YORK HOMŒOPATHIC COLLEGE, FORMERLY PROFESSOR OF PHYSIOLOGY IN THE NEW-YORK MEDICAL COLLEGE, AND HOSPITAL FOR WOMEN, FORMERLY PROFESSOR OF PHYSIOLOGY IN THE NEW-YORK HOMŒOPATHIC MEDICAL COLLEGE, MEMBER OF THE AMERICAN INSTITUTE OF HOMŒOPATHY, FORMERLY PRESIDENT OF THE AMERICAN HOMŒOPATHIC OPHTHALMOLOGICAL AND OTOLOGICAL SOCIETY.

BOSTON:

OTIS CLAPP AND SON,

3 BEACON STREET.

1885.

TO

D. B. ST. JOHN ROOSA, A.M., M.D.,

PROFESSOR OF DISEASES OF THE EYE AND EAR IN THE UNIVERSITY OF THE CITY OF NEW YORK, MY TEACHER THERE IN OTHTHALMOLOGY AND OTOLOGY, ALSO MY KIND ADVISER SINCE THEN;

AND TO

TIMOTHY F. ALLEN, A.M., M.D.,

PROFESSOR OF MATERIA MEDICA AND THERAPEUTICS IN THE NEW-YORK HOMŒOPATHIC MEDICAL COLLEGE, MY TEACHER AND EARLY GUIDE TO A SCIENTIFIC STUDY OF REMEDIES FOR THE EYE AND EAR, MY BELOVED FRIEND AND COLLEAGUE,

This Work is Dedicated,

IN THE HOPE OF THE ABOLITION OF ALL DIVISION-WALLS BETWEEN EDUCATED PHYSICIANS AND SURGEONS.

HENRY C. HOUGHTON.

PREFACE.

"Of making many books there is no end; and much study is a weariness of the flesh." Why add another book to the list? Because it was solicited. Be assured it would not have been added if not solicited. When the New-York Ophthalmic Hospital was transferred to the service of a different surgical staff, the course of instruction that had been established was continued. It devolved on me to give the lectures on diseases of the ear, and a written petition from one of the classes urged me to write a work which should embody the results of experience obtained in the hospital clinics. Such a work was under way in 1876, when serious, and at one time apparently hopeless, illness prevented its completion. In 1881 the manuscript was nearly complete, when the announcement of "The Human Ear and its Diseases," by W. H. Winslow, M.D., precluded, to my mind, the issue of my work.

Since then, my duties as Professor of Clinical Otology in the New-York Homœopathic College have given another occasion for a similar request for the substance, in book-form, of the lectures given to the senior class. The manuscript has been re-written from the notes of a stenographer; and no alterations have been made, except of errors in the structure of sentences, such as are liable to occur in extemporaneous speaking.

The book is not written for the specialist, but for the student and the busy practitioner, who will find in it suggestions for the treatment of aural diseases, and indications for remedies that have proved effective in a large clinical practice. At the request of some of my colleagues, there are added cases from private practice, which illustrate the action of certain remedies.

The repertory which is added to the lectures, is made from material furnished by my friends Drs. Hering and Allen. That it is my ideal, is not claimed ; that it may prove a stepping-stone to an ideal, may not be a vain hope. The winnowing of our materia medica may seem an easy task to the young, enthusiastic student ; but, the longer one labors at the task, the nearer he carries his sieves to the borders of the Slough of Despond. Some material has accumulated for a work similar to Allen and Norton's work in the field of ophthalmic therapeutics ; but this work is even more difficult, from the fact that aural pathology and diagnoses is a more unsettled field of investigation. To such work, my mind has been specially drawn ; and it is my purpose to preserve every reliable result obtained. Crude, unreliable results are worse than useless, serving as stumbling-blocks in our path.

To my colleagues, my effort is commended for their fraternal criticism, kind or caustic, only so it is true. If you find confirmations, please report them to me ; if errors, note them also, that we may have the truth. If any fellow-creature shall be placed in better enjoyment of special function, and thereby made happier and more useful, my hope will be realized.

The chromo-lithographs are reproduced from the plates of

Dr. Adam Politzer's "Monograph on the Membrana Tympani," issued by Messrs. William Wood & Co. in 1869, to whom I am indebted for permission to reproduce the same, as also for the electrotypes furnished by them at the written request of Professor D. B. St. John Roosa, M.D., from his work, "The Diseases of the Ear." It gives me great pleasure to acknowledge this courtesy of my esteemed instructor.

The cuts of instruments are furnished by Messrs. Meyrowitz Bros. of this city.

NEW YORK, March 3, 1885.

CONTENTS.

FIRST LECTURE.

SECOND LECTURE.

THIRD LECTURE.

FOURTH LECTURE.

FIFTH LECTURE.

SIXTH LECTURE.

SEVENTH LECTURE.

EIGHTH LECTURE.

NINTH LECTURE,

TENTH LECTURE.

ELEVENTH LECTURE.

TWELFTH LECTURE.

APPENDIX.

LIST OF ILLUSTRATIONS.

PART FIRST.

DISEASES OF THE EXTERNAL EAR.

FIRST LECTURE.

INTRODUCTION.

Gentlemen, — It gives me pleasure to meet you in this place, not unfamiliar to me in my former relation as professor of physiology in this college. My present position, however, may be termed a new departure; and it may not be unfitting to refer to the providence which has led to this position.

At the close of the late war I came to this city, from my home in Boston, Mass., with letters to the President of the Medical Department of the University, and to Professor A. C. Post, M.D. The latter gentleman received me kindly, and, as his own clinical staff was full, gave me a letter of introduction to Professor D. B. St. John Roosa, M.D., who kindly invited me to a place on his staff as clinical assistant, which position I held during two sessions. The acquaintance thus begun has continued to this day.

Later, while physician to the "House of Industry," I was favored with the advice and counsel of the late Carroll Dunham, M.D., and Professor T. F. Allen, M.D.: to the latter gentleman I am more especially indebted for initiation and later guidance in the study and practice of the principles of homœopathy. In 1868 the "New-York Ophthalmic Hospital" was placed in charge of a surgical staff, the members of which were practitioners of homœopathy. In December of that year I received the appointment of aural surgeon. During the first year the number of aural patients treated was fifty-seven, I think; and within seven years the number rose to nearly one thousand. From the experience thus gained, I hope to be able to direct you to the methods of instrumental and medicinal treatment which in my practice have been proved efficient. It is my purpose in these lectures, to present to you, in connection with such clinical material as shall offer, the general features of diseases of the ear; and I shall endeavor to present them in a way so practical that the suggestions which you receive, may prove of value in your hours of need as busy practitioners.

The motives which impel one to study are usually mixed; and, while the greater should always be a desire to benefit one's kind, the secondary one, of personal benefit, is not unworthy of notice.

You will find in experience, that knowledge and skill in the treatment of special diseases will bring legitimate practice in other directions Certain facts may induce you to give more careful attention to these diseases, and impress you with their importance. We have to consider their frequency, their serious nature, their effect upon life and usefulness, also their agency as a direct cause of death, their anatomical relations, giving rise to cerebral disease.

First, As regards their frequency. In 1856 Kramer reported, that, in Prussia, statistics showed one deaf-mute in 1,109, and one blind person in 1,730. If we add to the deaf-mutes the number of those patients affected with other ear-diseases, we shall see that the difference between the two is even greater.

Notice, secondly, the serious nature of these diseases in their effect upon human life, determining, not only its limit, but its usefulness. The child that loses his hearing before seven years of age, may become a deaf-mute; and, if not a deaf-mute, the limitation put upon this avenue to the brain dwarfs his whole mental development. Then, too, the child that has suppurative disease is often the object of aversion, becomes morose, misunderstands others, is misunderstood in turn, and grows up under this acquired antagonism to his fellows.

It is now known that the causes of death in many cases

hitherto mysterious, are not lesions of the heart or of the spinal cord, but of the brain, and result directly from abscesses, which are secondary to disease of the ear. Notice now the anatomical relations. Above, the cerebrum, separated only by a thin plate of bone; below, the maxillary articulation, jugular veins, carotid artery, and numerous nerves; posteriorly, the mastoid cells in immediate relation with the cerebellum. Suppuration occurring here may cause mischief in either of these directions. In fact, in the child, before the temporal bone is fully formed, the relation is direct to the cerebrum; while, in the adult, it is more direct to the cerebellum. It is a fact that no branch of surgical practice has so failed from lack of general interest as aural surgery. The reason is found in the apathy of the profession toward it, which has been due, undoubtedly, to the earlier fact that this branch of practice was in the hands of charlatans; relegated even to barbers. To-day there is no reason for lack of interest: no department of medical or surgical study has shown such advance as aural surgery during the last quarter of a century. This advance is due to pathological, as well as to clinical, study. First to provoke an interest in pathological research is to be named Toynbee, the English aurist, who has left behind him, in "Hunter's Museum of the College of Surgeons," London, a collection of morbid specimens which is a monument to his patience and skill. In later years, English, Continental, and American surgeons have added their quota to the store of knowledge.

This has been supplemented by clinical study, both abroad and in this country, to such an extent that we are left without excuse for any neglect in the care of this class of patients.

Turning, now, to the subject which shall engage us, I ask you to notice the contrast between the eye and ear as regards clinical instruction. Very many diseases of the eye can be studied without instrumental examination; but, in the study of the diseases of the ear, in almost all cases the use of instruments is required.

As regards the instruments, I shall simply call your attention, at present, to those more easily employed. To the use of the aural mirror is due most of the advance made in clinical otology. It was invented by Von Troltsch immediately

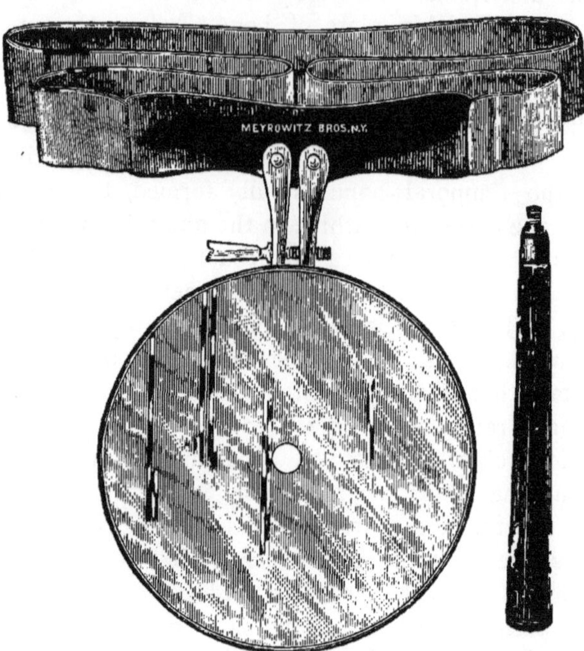

MIRROR AND HEAD-BAND.

after the introduction of the ophthalmoscope, and involves the same principles, modified for its special uses. Previously

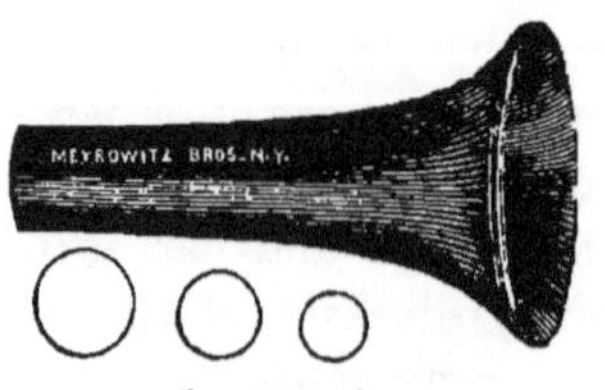

TOYNBEE'S SPECULUM.

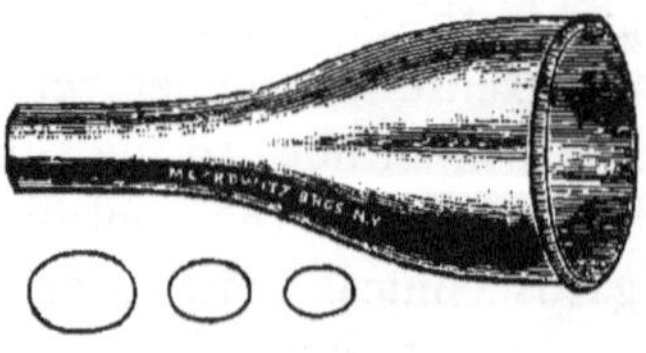

GRUBER'S SPECULUM.

to the introduction of this instrument, the surgeon depended upon direct illumination, which was almost futile on account of the narrowness and depth of the auditory canal. It now supersedes every other means of illumination. In connection with this instrument we have the aural speculum, modified according to the ideas of different teachers. The one which I use, and prefer, is Toynbee's, which, having an

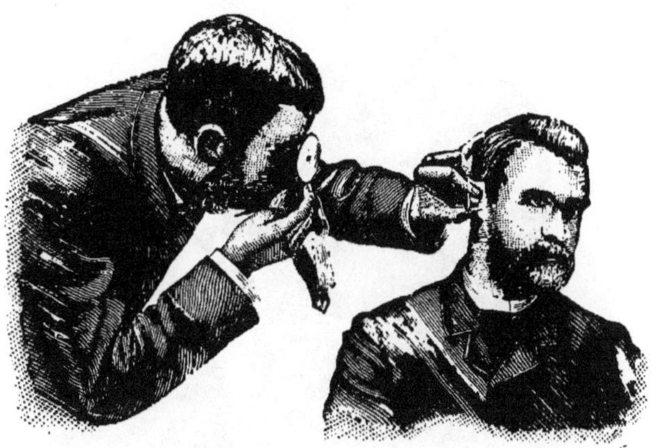

METHOD OF EXAMINING EXTERNAL AUDITORY CANAL. (ROOSA.)

oval opening, can be adapted to the long diameter of the auditory canal, giving us a more complete view of the membrana tympani.

The Eustachian catheter, Politzer's apparatus for inflation, and Siegel's pneumatic otoscope, are here shown you. The method of their use will be demonstrated in connection with cases, but I will say a few words as to their application in diagnosis as well as treatment. The first is of special value in cases in which only one ear is affected, as inflation by those methods that affect both ears may act unfavorably on the healthy ear. With the catheter the air can be forced directly into the middle ear; but its use is open to the objection, that it causes irritation of the mucous membrane of the nares, and mouth of the Eustachian tube, and in care-

EUSTACHIAN CATHETER.

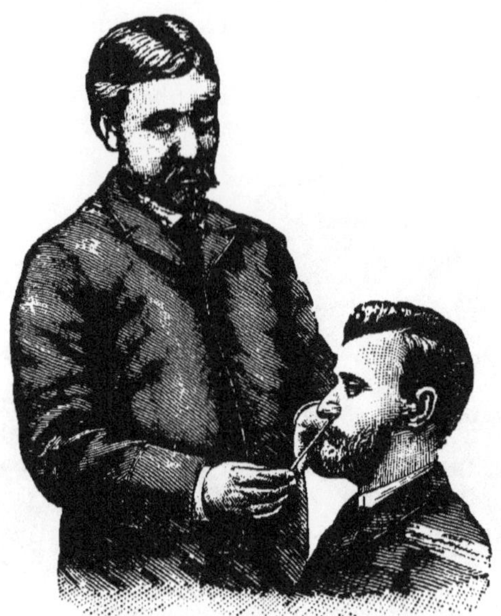

INTRODUCTION OF EUSTACHIAN CATHETER. (ROOSA.)

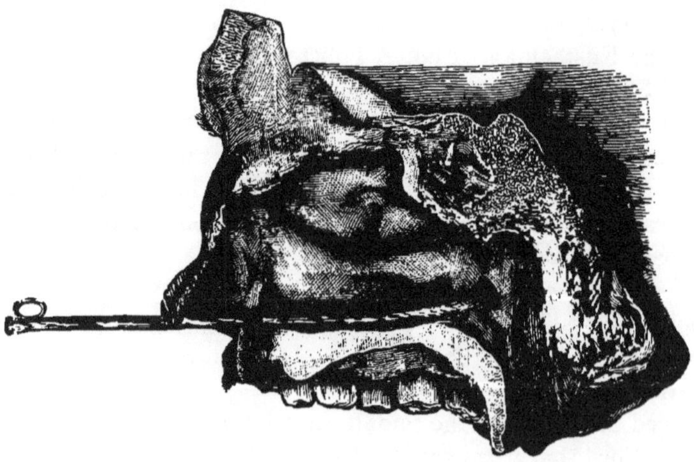

SECTION SHOWING MOUTH OF EUSTACHIAN TUBE. (ROOSA.)

less hands has done mischief; but it is not exceptional in that respect. The introduction of the catheter is an easy matter in most cases. Dexterity comes by practice. Hold the instrument free in the fingers, not stiffly, thus: introduce the beak while the instrument is almost perpendicular; raise slowly to the horizontal, pass slowly back, withdraw slightly if you meet obstruction; turn a little to one side, pass back till the beak is free in the posterior nares, press back till the instrument is stopped by the posterior wall, then draw slightly forward, turning the ring of the instrument toward the ear affected. If the beak drop into the ostium of the Eustachian tube, it will not move during the act of deglutition; if it do so, it must rest on the lateral wall; turn the

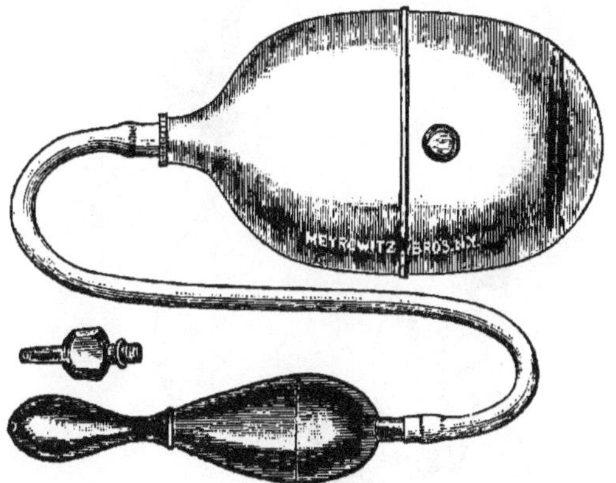

POLITZER'S APPARATUS.

ring to its first position, and repeat the manipulation; a little patience, and it will come to its place; now use the air-balloon, and the air is felt in the middle ear.

Politzer's method of inflation consists in the use of the air-bag, with a flexible tube, having a suitable nose-tip, which is introduced in the nostril, the opposite one being closed by the fingers while holding the nose-tip in position: the patient is directed to blow as in putting out a light, or pronounce hock; and at that moment the air is forced from the air-bag by sudden closure of the hand. The pharynx being divided, by the action of the soft palate, into superior and inferior portions, the openings of the superior portion are the Eustachian tubes and the nasal passages. Now, if the latter are closed, and air be forced through the nose-tip as above, the air must pass into the tympanum. If whistling or articulation fail to cause strong action of the pharyngeal

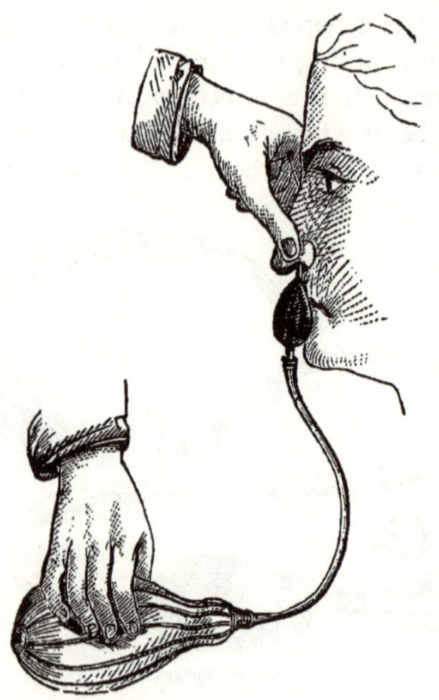

Method of using Politzer's Apparatus. (Roosa.)

muscles, direct the patient to hold a little water in the mouth, and, upon commanding him to swallow, force the air from the air-bag as the water passes from the mouth. If this fail, use the catheter, or give the remedy best indicated in the case, and wait its action for a few days, till the condition of the mucous membranes is changed ; then inflation will be accomplished more readily ; indeed, in many cases, the remedy causes the tube to open spontaneously. This method is free from the objection urged against the catheter, but is open to another: frequent and forcible use causes distention of the parts, and injury has resulted from trusting the instrument to lay hands. The same may be said of Valsalva's method, or rather experiment as it is called. This is performed by closing the nares with the fingers, and attempting to blow through the nose. This method of inflation should never be used as a means of treatment, and patients should be informed of the bad results which follow its practice. We often see relaxed and flaccid membrana in the persons of those who have learned to "blow out the ears."

Siegel's otoscope is very valuable for settling various points on differential diagnosis, — such as conditions of the external layer of the drum-membrane, perforations in the

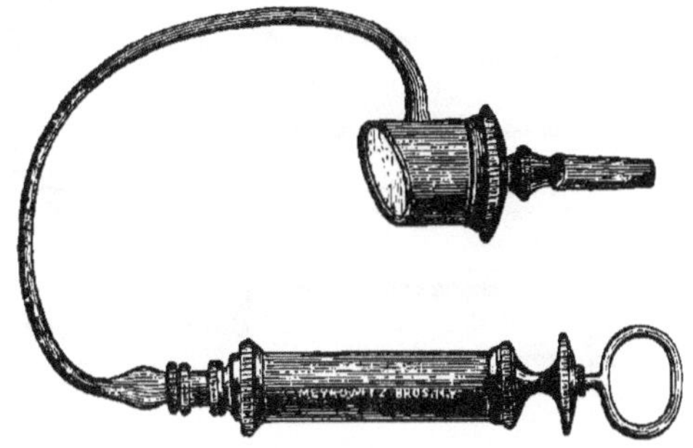

SIEGEL'S OTOSCOPE.

same, adhesions in the tympanum, etc. It is also applicable as a means of producing passive motion in the articulations, and, possibly, of breaking adhesions between the membrana tympani and the ossicula or the labyrinthine wall. By introducing the speculum firmly in the meatus, and alternately forcing the air in, or exhausting it, the movements of the drum-head may be studied under strong light.

The stop-watch[1] is of great value in testing the hearing-distance; as by it you are able to exclude any illusions on the

[1] The Auburndale timer, sold by Cross & Beguelin, No. 21 Maiden Lane, New York, is well adapted to the purpose, and is not expensive. There is also a similar timer, made in Switzerland, which can be bought in this city at less price.

part of the patient, by alternately starting and stopping its action. The distance at which the watch is heard should be recorded in fractional form, as suggested by Dr. Prout of Brooklyn, and modified by Drs. Knapp and Roosa. H, represents the watch; V, spoken voice; v, whispered voice. Hence, H 20-20 means normal hearing; 10-20, a watch heard by person with normal hearing 20 feet, is heard only 10 feet. If heard only 1 inch, the expression is 1-20×12=1-240. The record for spoken voice thus: V 60-60; that is, the normal hearing is 60 feet; then, by 10-60, we understand the voice is heard by the deaf person, under examination, only 10 feet.

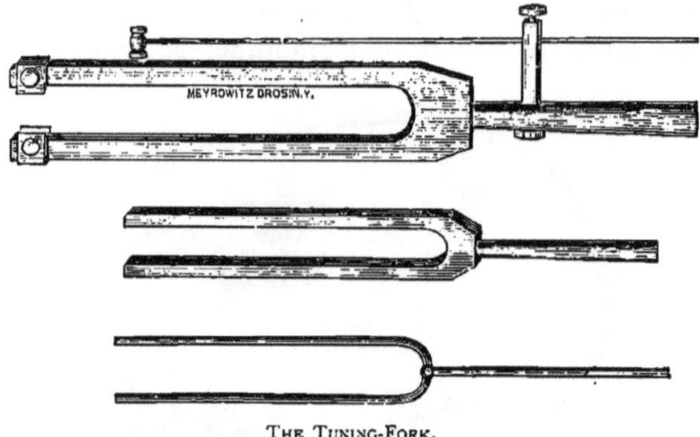

THE TUNING-FORK.

The tuning-fork is used to determine the state of the acoustic nerve, or to differentiate between lesion of the internal and middle ear. In the normal ear, the tuning-fork, vibrating before the ear, is heard better through the air than when placed on the bones of the skull. In disease, if the fork is heard better through the air than when placed on the skull, it is evidence that the acoustic nerve is affected: but, if the fork is heard better on the bones, it indicates disease of the middle ear, or external auditory canal, or both; because rigidity of the ossicula, or obstruction of the meatus, causes increased and prolonged vibration of the auditory nerve.

Care should be taken to make complete and full record of each case at the first, and on each subsequent, visit. By comparing these records, you are better able to judge of results; and they will prove invaluable to you as matters of history and statistics.

In the course of these lectures, I shall endeavor to follow, as closely as possible, the classification and nomenclature which I shall give you; but it will be necessary to deviate from them, for material offered at the clinic is not always classified to our hand.

Diseases of the ear are divided according to their anatomical basis, — otitis externa, media interna. Otitis externa embraces all the diseases of the auricle, lesions of the dermoid structure, of cartilage, and of the external surface of the membrana tympani. Under these may be named, diffuse inflammation, either acute or chronic; circumscribed inflammation, always acute; myringitis, ulcerations, necrosis, polypoid growths, parasitic diseases, exostosis.

NOTE. — In using Politzer's method of inflation, I have recently found, that directing the patient to blow through a small glass pipette gives better results than either of the methods formerly used. The patient is directed to hold the tube in the hand opposite the side to be treated; place it in the corner of the mouth; take a full inspiration, and blow with great force; at the moment of greatest effort, use the air-bag as usual.

SECOND LECTURE.

CIRCUMSCRIBED INFLAMMATION AND DIFFUSE INFLAMMATION.

GENTLEMEN, — *Otitis externa*, or inflammation of the external ear, is divided into circumscribed and diffuse.

Otitis externa circumscripta (peri-follicular inflammation) is inflammation of the dermoid elements and connective tissue of the external third of the auditory canal, commonly known as furuncle.

Symptoms. — The earliest symptoms consist of an itching or burning sensation, which later increases to actual pain. This pain is not usually severe, unless the tissues beyond the outer third are involved. If this be so, as is the case when the patient is suffering from a severe series of furuncles, the pain is agonizing, associated with great constitutional disturbance, and in some cases with delirium. The ordinary experience, however, is possibly twenty-four hours of suffering, followed by a lymphy or purulent discharge from the meatus; and the disturbance is passed for the time. The objective symptoms are swelling, redness, and, on touching with the probe at some circumscribed spot, the patient will shrink on account of the pain. The history of the trouble is usually short, the resolution prompt. After a few days the dermoid structures are exfoliated, and the meatus assumes its normal appearance.

Cause. — The cause is usually to be found in the direction of excess, or in lack of proper diet, or some wrong habit which modifies nutrition; so that we may say, furuncle is the rich man's penalty or the poor man's necessity. The fact that furuncles occur either in spring or fall may suggest a constitutional change incident to the season. Such symptoms demand attention.

Treatment. — The treatment consists in the local application of some remedy to mitigate the pain, and the administration of such remedies as will overcome the habit. *Plantago major*, fluid extract, and warm water in equal parts, dropped into the meatus as warm as can be tolerated, or applied on absorbent cotton, or a few drops of Magendie's solution applied in the same way, will ameliorate the pain, and will not interfere with the action of internal remedies. Among the various remedies that have been used are *belladonna*, *pulsatilla*, *hepar sulph.*, and *mercurius*.

As regards the incision of furuncle, I am satisfied that it is better practice not to cut deeply, to abort, or to evacuate, the abscess, but to allow it to take its course; as the worst cases of repeated inflammation that have come under my observation have been those in which incisions were made. Soon after the proving of picric acid, my attention was called to that remedy by Dr. Clara C. Plympton, a graduate of the Ophthalmic Hospital; and since that time the remedy has been used for that disease, and may be considered as nearly a specific for it as is any remedy for any disease. Last spring I had occasion to prescribe *picric acid* in the case of a young lady, twenty-five years of age, who had suffered from a series of furuncles in the meatus, which bade fair to reach the mystic number seven, — this being the limit which in the lay mind is supposed to exist. Picric acid relieved the symptoms, but *hepar sulph. calc.* was necessary to complete the cure. A relapse and the necessity for a repetition of the picric acid suggested to me a combination of these two drugs. I therefore obtained through Messrs. Boericke & Tafel the compound salt: it is called *calcarea picrata*, and it has been used in a number of cases in the hospital during this session. It bids fair to warrant the late Dr. Hering's statement, that the combination of two remedies in a compound gives the curative action of each in a greater degree.

Otitis externa diffusa may be divided into two forms, acute and chronic.

Cause. — The acute form usually results from some external irritation of the tissues of the meatus, arising from the use of some instrument in picking, scratching, or digging into the meatus, as is the habit of many persons.

Symptoms. — The symptoms are itching, burning, with a sensation of obstruction in the meatus, and a greater or less degree of loss of hearing. The chief objective symptoms are closure of the meatus, and the oozing of lymphy or purulent fluid. On touch with the probe, pain is caused in all parts of the meatus, in contrast to *otitis externa circumscripta*, in which pain is caused only on touch of the furuncle. The course of the disease is less rapid than the circumscribed form, often lasting a number of days or weeks, and tends to a chronic type. If its course be limited, and resolution be prompt, the dermoid structure of the meatus will be exfoliated, and, in many cases, can be removed almost entire: on removal, the outer third of the canal will be left denuded, red and shining. This process of exfoliation will be repeated until the integument assumes its normal condition.

Treatment. — The suggestion previously made regarding the use of some remedy to mitigate the pain, holds good in this form of the disease. You will understand, therefore, that I am not an extremist in the administration of internal remedies, but believe that the mitigation of symptoms by local application is judicious practice, assured as I am that it does not interfere with the action of internal remedies given.

The remedies for the acute stage are, *aconite* in a few instances; more usually *belladonna, ferrum phos., hepar sulph., mercurius, kali mur.*, and *pulsatilla.* The differential diagnosis between *aconite* and *belladonna* will be clear to your minds at the sight of the extreme restlessness, flushed face, and great thirst of aconite patient. The choice between *hepar sulph.* and *mercurius* is, that the hepar is indicated in extreme sensitiveness to the air, — which is relieved by wrapping, — aggravation at night. Sensitiveness to touch is more localized than in mercurius. In the latter remedy, the nightly aggravation is more marked than hepar sulph., the sensitiveness to air is less, the soreness more generally about the ear. *Pulsatilla* is more frequently indicated in circumscribed inflammation, but in women and children its characteristic symptoms will sometimes occur. *Chamomilla* will be indicated in some few cases by the extreme tolerance of pain.

Instrumental treatment is more likely to be required than in the circumscribed form; as it will be necessary to remove all portions of exfoliated epithelial structures from the canal, lest they become the nidus of plugs of detritus, of cerumen, etc. The application of fluid petroleum with a camel's-hair brush will relieve the intense irritation and itching which mark the process of resolution.

Associated with this strictly technical division of our subject is the matter of foreign bodies, mal-secretion of cerumen, skin-diseases of the auricle, and region above the ear, parasitic growths, as well as animal parasites.

Eczema of the auricle, or the region immediately about it, is the same here as in other parts of the body; and there is nothing special to say in regard to its treatment other than that *arsenic, graphites, mezereum*, and *sulphur* are the remedies most frequently indicated. The deep cracks which form behind and below the auricle, yield to graphites; and we have found the cure to be hastened by the local application of graphites combined with concentrated petroleum, which has been largely used in similar forms of inflammation of the eyelids.

Psorinum is an invaluable remedy, often superior to sulphur in cases in which the latter remedy has been formerly used.

The following symptoms, as given by Hering, we have confirmed again and again. External ears, raw, red oozing; scabs form; sore pain behind the ears. Otorrhœa very offensive, purulent; watery, stinking diarrhœa. Pustules on and behind concha. Scabby eczema behind right ear. Herpes from temples over ears to cheeks at times throws off innumerable scales; at other times, painful rhagades with yellow discharge, forming scurfs, fetid humor; itching intolerable.

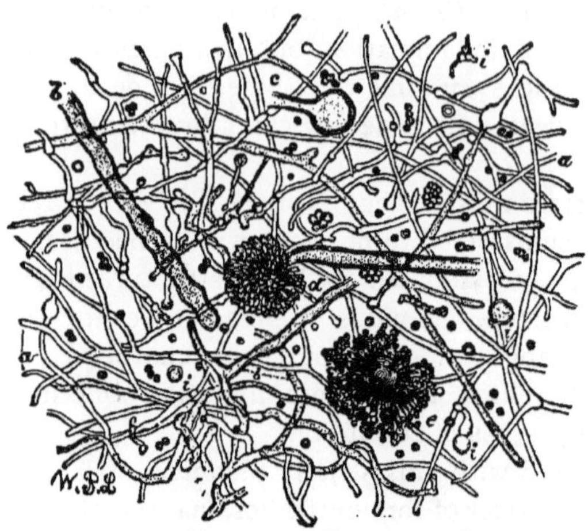

ASPERGILLUS NIGRICANS, 220 DIAMETERS. (ROOSA.)

a. Mycelium fibre. *b.* Fruit-bearing fibre. *c.* Naked sporangium. *d.* Sporangium covered with basidia only. *e.* More mature sporangium. *i.* Spores in a state of germination.

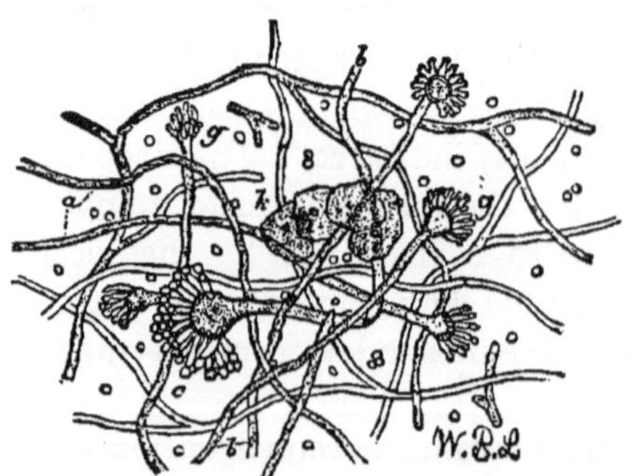

ASPERGILLUS FLAVESCENS, 220 DIAMETERS. (ROOSA.)

a. Mycelium fibre. *b.* Fruit-bearing fibre. *c.* Sporangium-bearing spores upon the basidia *g.* Basidia showing construction preparatory to the separation of spores. *k.* Epithelium.

Hæmatoma is most satisfactorily treated by evacuation of the contents of the sac, and application afterwards of pressure. In a few cases, the galvanic cautery has proved efficient in forming clot, and possibly hastening the resolution after the evacuation and pressure.

Malformation of the auricle or of the lobule, or abscess, and cutting of the ring-hole and lobule by heavy earrings, are the subjects of plastic operations, which belong as much to the sphere of general plastic surgery as to aural surgery.

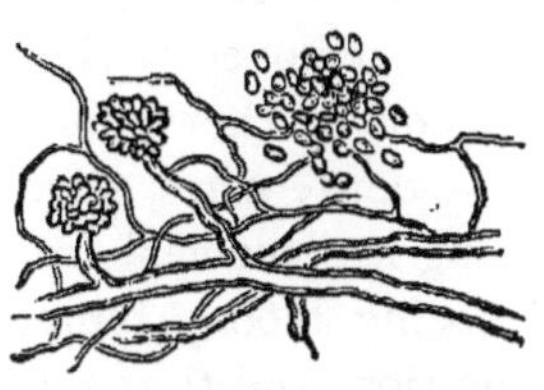

PENCILLIUM GLAUCUM. (ROOSA.)

Otitis externa parasitica is caused by the presence and production of microscopic vegetable parasites. The most usual form is that of aspergillus glaucus, which is interchangeable with pencillium glaucum. There was a difference of opinion concerning the production of this parasitic growth, — some claiming that a spore being deposited in the canal caused the inflammatory action, and flow of lymphy pus; others, that a previous erosion, abrasion, or maceration of the tissue gave the necessary moisture which facilitated the growth of the fungus. The latter view now prevails. The growth consists of mycelium, which forms the bed, or mass, of the growth, from which rise the small trunks: on the tops of these trunks are produced the bundles of spores, which, when ripe, are shed beneath, forming new centres of growth. In this way the mass is reproduced until it covers the entire walls, layer upon layer on the surface, and the whole canal is occluded.

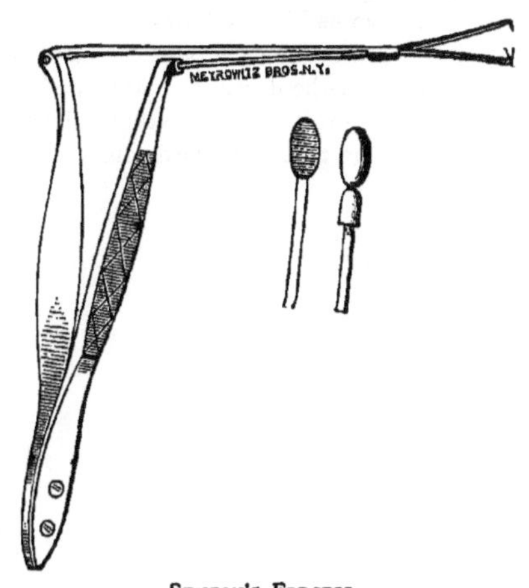

SEXTON'S FORCEPS.

Symptoms. — The symptoms are obstruction, with a low

degree of pain when the occlusion of the meatus becomes complete; and, in extreme cases, pressure upon the canal will cause very severe reflex symptoms in the head, even as far as the neck and arm. The chief symptoms are similar to those caused by a plug of cerumen, with this difference, however, that ordinarily the cerumen will be removed without difficulty, and without causing any abrasion of the walls of the canal: the parasitic growth forms a dense mass, which one is unable to remove except by tearing it piecemeal from the walls, leaving them abraded, and oozing bloody lymph. The whole canal will be left in this condition, and thus form a very favorable field for the reproduction of this growth, unless some agent is used to destroy these microscopic organisms.

NOTE.—I have utilized the Eustachian catheter and the ear-syringe together in a way that I have not seen advised by another. Instrument-dealers make the syringe with a blunt tip and a posterior-nares tip. These are interchangeable. Messrs. Meyrowitz Bros. have added to these, three Eustachian catheters. The advantage gained by the latter is in the fact, that the surgeon can get an unobstructed view of the meatus, as the hand is turned to the side, right or left, above or below, while he directs the stream of water in any direction. Another advantage gained is this, that the catheter does not interfere with the outward flow of the water, and cause awkward spills.

THIRD LECTURE.

ACCUMULATION OF CERUMEN. — REMOVAL OF FOREIGN BODIES.

Gentlemen, — We will take first to-day as the subject of our consideration, the *Accumulation of Cerumen.* There is no reason why the glands of the skin of this locality should not be subject to deviation from normal action; and we find deviations in both directions, excessive action as well as lack of proper secretion.

The symptoms of impacted cerumen are, sudden loss of hearing, tinnitus aurium, vertigo, sense of fulness. A mass of soft cerumen with the slightest chink between it and the wall of the meatus will cause no trouble; but when under any circumstances it is dislodged, the mass will produce at once the foregoing symptoms. Von Troltsch mentions the case of an old man, who, on his way home after a wine-supper, fell over the pole of a wagon. For fifteen minutes he was apparently senseless, but recovered sufficiently to reach home. The next morning his physician diagnosed the case as concussion, or possibly apoplexy, due to the fall, or blow on the head from striking the pavement. The patient, who was otherwise well, was dieted, cupped, purged, and, after a few days, a seton was introduced into the back of his neck. A month later Von Troltsch was called, and found both canals stopped with cerumen. He was immediately relieved of deafness, and of the "profound cloudiness of intellect" following concussion of the brain. Roosa mentions a similar case, which was attributed to sunstroke.

Causes. — As has been intimated, the accumulation is due to the excessive action of the sudoriferous and ceruminous glands; the immediate cause being efforts at cleanliness, or manipulation with various instruments, to relieve the ear from symptoms that are caused by conditions of the middle ear.

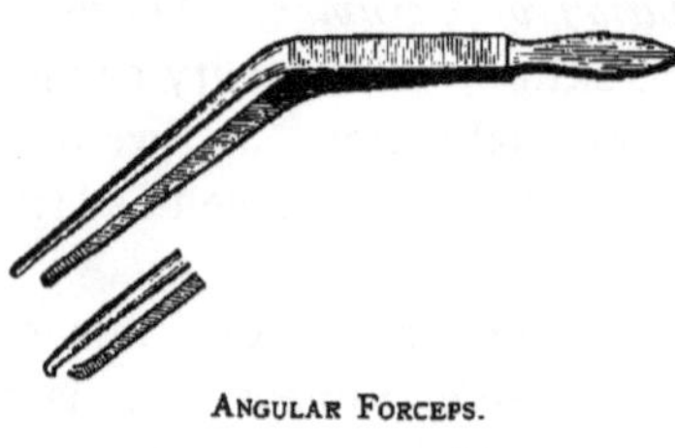
Angular Forceps.

Treatment. — In all cases, examine the ear with reflected light. When the canal is well illuminated, move the mass, if possible, so as to form a passage on either side along the wall of the canal. The best instrument for this is Buck's loop-probe, or *curette.* Then with a syringe direct a stream of water into the passage, or chink, which has been made between the mass and the wall of the canal. In this way, you will often dislodge the entire mass by one or two efforts. In some cases, you will be unable to remove the mass at the first sitting without causing more irritation than is safe. In such a case, direct the patient to instil into the meatus every night five drops of fluid *petroleum*, or of a mixture of three drachms of *petroleum*

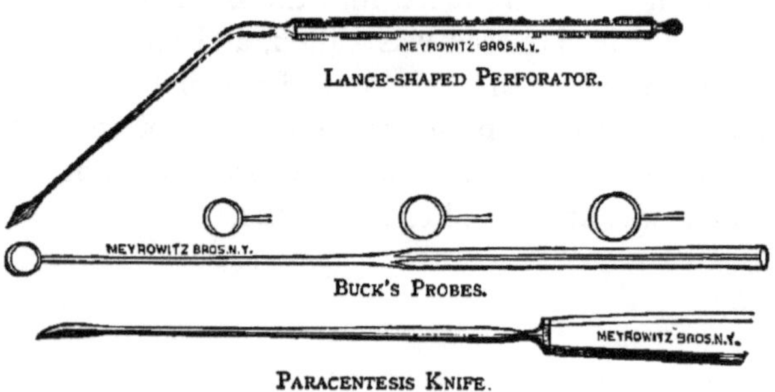

Lance-shaped Perforator.

Buck's Probes.

Paracentesis Knife.

and one drachm of *sulphuric ether.* The effect of either of these is to concentrate the mass, to separate it from the walls; and you are often able to remove easily the entire mass with forceps after a few days.

A peculiar condition is sometimes noticed, in which the cerumen, besides forming in thin scales along the walls of the meatus, covers the membrana tympani with similar scales, and interferes decidedly with the function of hearing. The instilled petroleum will separate such scales, which can easily be removed with cotton upon the end of a proper holder.

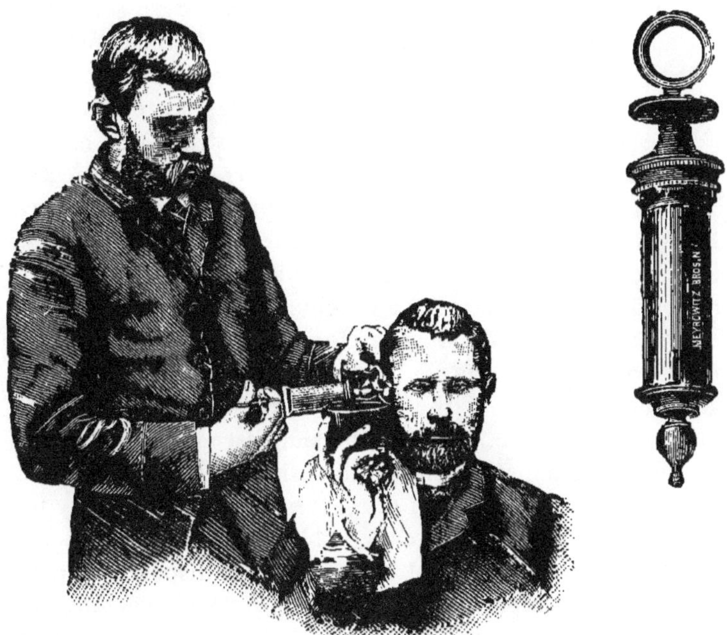

METHOD OF SYRINGING THE EAR. (ROOSA.)

In some cases, there is noticed a condition called keratosis obdurans. Opinions differ as to the nature and formation of this mass; but I believe that the weight of authority is in favor of the view, that it is simply a mass of cerumen and scales, which has become impacted by slow accumulation and long exposure till it has become hornlike. In such cases, the mass is not usually removed entire, but is first broken into small pieces by the probe, and then removed by forceps.

The conditions which exist after the removal of these masses often require attention. The pressure of the mass may give rise to circumscribed ulcerations: these being neglected would become the foundation of future similar troubles. I would suggest pencilling the canal with petroleum combined with calendula, and occasional inspection until the normal condition is observed. Be careful always to test the hearing after the removal of foreign bodies, so that, in the great relief which follows the removal, you may not neglect the diagnosis of an existing middle-ear disease.

Foreign bodies. — At the close of my last lecture, I found a little patient whose case would have illustrated the method of removing small foreign bodies from the external meatus. The child, while playing with some grain, had put several kernels in the ears. A day or two later, the parents learned this fact, as the grain caused some annoyance; attempts at removal only aggravated the mischief: and the child resisted examination, fearing more suffering. Illumination of the canal without the speculum showed the end of one kernel, and one discharge of the syringe filled with warm water removed all the grain. This is the simplest method of procedure when examination shows that the stream of water will pass between the walls of the external meatus and any foreign body, such as bits of cork, pebbles, beads, etc., — in fact, any thing that the "awful boy" may feel disposed to place within his meatus. The same suggestions touching the removal of hard masses of cerumen hold good in these cases; and your own surgical skill, enforced by clever wit, will help you out in many instances by peculiar devices. In a case mentioned to me by the late Dr. Carroll Dunham, a lad plugged his ear with a cork from a vial, completely occluding the meatus. Dr. Dunham used a dental excavator; and, by patiently excavating a triangular cavity with its base inward, he was able to pass an explorer into the cavity, securing its point in the side, and drew the cork from the ear. Lowenberg mentions the application of a camel's-hair pencil, saturated with glue, to the surface of an ivory ball, leaving it until it adhered perfectly, then withdrawing pencil and ball at one effort. A writer suggests the use of loops of horsehair, three or four hairs gathered together, both ends in the fingers, pushed into the ear as far as the drum-head, then twisted in the fingers, to remove a small foreign body. Another suggests the expedient of turning the ear involved towards the floor, as the patient lies on the table; then, as the head projects beyond the table, syringe from below, upward, and the weight of the body will aid in its fall from the ear. I have succeeded in removing a number of foreign bodies, which have by previous manipulation become impacted, by using a probe bent at right angles, similar to a tenaculum, pushing it between the body and the wall while the patient was under ether. Having passed the foreign body, turn the tenaculum at right angles to the wall, and, on withdrawing the instrument, the foreign body must necessarily come with it. Animate foreign bodies, as flies, bed-bugs, croton-bugs, have all been subjects of our attention at the clinic. In two cases the cimex had entered in an animate state, and yielded in one case to warm oil, and, in the other case, to a dose of brandy and water; the sufferer being a denizen of the region of the "Five Points," this remedy was probably a panacea for all his troubles. The simplest method is to instil warm water or warm oil; and, later, the surgeon can easily remove the insect by the use of the syringe.

Writers mention cases in which insects have found in the secretion of the ear a favorable nest for their eggs, and, later, larvæ in large quantities have developed. The use of warm water with a few drops of carbolic acid, or an alcoholic solution of boracic acid, will dispose of the whole matter.

The laity, and even some members of the profession, have the idea that a foreign body in the meatus is a matter of great danger. It is not so. The immediate danger is less grave than dangerous interference: hence be sure you see the body before you attempt its removal. Harsh attempts at removal cause impaction. Serious consequences have followed ill-advised attempts at removal. One standard authority reports death as the result of efforts, with probes and forceps, to remove a nail which was supposed to have lodged in the meatus; and after the removal of several small bones, shreds of membrane, etc., the patient was left in collapse, on the supposition that the nail had passed into the cerebral cavity. The *post-mortem* demonstrated the absence of any foreign body.

The presence of a foreign body in the meatus will give rise to peculiar and seemingly grave symptoms in some cases. Some time ago a member of the fire-department called at the clinic, saying he had let water in his ear while bathing his head at the hydrant. The full stream had struck him on the head. Soon after, an irritation of the ear, neck, and shoulder had set in, alarming the patient, who feared paralysis would result. On examination, a white mass was seen at the lower inner extremity of the canal, resting against the drum-head. Syringing with warm water brought away a quartz pebble, and all irritation of the nerves on that side of the body ceased at once.

We come now to consider some lesions of the membrana tympani. This structure is so important in relation to the objective symptoms, that more than a passing notice is demanded. From its relation to the external as well as the middle ear, it partakes of the nature of both, and is subject to changes having their beginnings either in the meatus auditorius externus, or in the cavity of the tympanum: so that Politzer divides the changes noticed in the membrane into softening of the epidermoid layer, thickening of the same, also thickening of the mucous layer; the first is due to the serous exudation in the meatus, and is as near an idiopathic myringitis as will be found. True myringitis is a rare disease of the membrane; being, as has been intimated, secondary to a chronic otitis externa, to a catarrhal, or to a suppurative, disease of the middle ear. In disease of the external ear, the membrane itself will be dull and thickened, so that the line of the bones will be obscured: whereas, in disease of the middle ear, the mucous surface being affected, will not interfere with reflected light, but will interfere with transmitted light; and, the translucency of the membrane being lost, the incus and stapes will be unseen. Opacities are noticed as the result of disease of the external layer. These are due to changes from suppuration or circumscribed ulceration in the membrane. Opacities of the mucous membrane give a bluish-gray or copper color to the membrana tympani, leaving it opaque, like ground glass. Circumscribed opacities, so-called calcareous deposits, are probably due to the disease of the mucous membrane, fatty degeneration, followed by change into amorphous calcarea carbonica. Tendinous or fibrous opacities cause very slight functional disturbance, and are of less consequence than adhesions between the membrana tympani and the tympanic wall, or between the membrane and the ossicula. This matter will be made clear when your attention is called to cases of perforation of the membrana tympani, or of adhesions between the membrana tympani and the ossicula, healing of perforations, cicatricial appearances, perforations of cicatrices, etc.

FRACTURE OF THE MALLEUS.

FRACTURE OF THE MALLEUS REPLACED. (ROOSA.)

The membrana tympani may be injured by explosions and by direct violence, by blows, introduction of instruments, etc.

Artillery-men have suffered from explosions. In most of these cases, it is true that there is an existing catarrh of the middle ear and Eustachian tube. Gruber has demonstrated that the healthy membrana tympani will sustain four or five times the atmospheric pressure. Examinations of workmen employed upon the foundation of the East-river Bridge, and also on the St. Louis Bridge, demonstrated the fact, that the laborers who were free from catarrhal diseases did not suffer any trouble in passing through the locks to the caissons, but those who were caught in the attempt at "changing the ears" (Valsalva's experiment) were subject to catarrhal diseases. Blows on the ears, as in the reprehensible practice of boxing the ears, may cause the rupture of the membrane, if it be softened by suppuration. Instances are recorded in which the membrane has been ruptured by twigs being forced into the meatus, and one case is mentioned of rupture by a pen-handle. Rupture of the membrana tympani by attempts at removal of foreign bodies, by excessive efforts in sea-bathing, and as a complication of phthisis, may come under our observation. Fracture of the handle of the malleus is reported by Wier.

Diseases of the External Ear. — Cases.

Mrs. John F. P., aged forty years. *Otitis externa circumscripta.* Nov. 19, 1880. Suffered for four weeks with cold. For more than a year has had itching in the ears; both canals so infiltrated and swollen, that only a partial view of meatus could be obtained; tissues of the canal infiltrated and oozing, but more sensitive at some points than at others. *Picric acid* three times a day, *belladonna* in water at night, or when the pain is more than usually severe. Under this treatment and local cleanliness, the tissues of the canal cleared up; but the habit of the tissues, which had been produced by prolonged fomentations, poultices, etc., was not readily overcome. Successive furuncles formed; and these were evacuated as rapidly as they pointed, no deep incision being made. From the very first exhibition of these remedies, the relief was marked; and the patient gained more in one week than she had previously done in four.

Dec. 4. The *salicylate of quinine* was given, before meals, to remedy the extreme prostration resulting from this prolonged tax upon the system.

Mrs. H. *Otitis externa diffusa acuta.* July 18, 1872. In right meatus, accumulation of scales and cerumen; canal sensitive; H. D. right, 3-240; left, 20-20; removed a portion, and instilled cosmoline.

July 20. Slight face-ache; H. D. 10-20; pain in mastoid process and on malar bone. *Capsicum.*

July 27. Pain again; slight amount of pus; H. D. 20-20; walls of canals normal. *Carbo veg.*

Nov. 20, 1873. Has a similar condition in right meatus; dermoid tissues exfoliating; inner wall red and tender; H. D. reduced to a minimum again; small polypus on the floor of the meatus. *Kali sulph.*

Nov. 21, 1874. Continued to improve until to-day; took cold; every beat of the heart is felt in the ear; *Mt.* bulged and red; very sensitive to air and touch. *Hepar sulph.*, with *belladonna* at night in water.

Nov. 26. Has improved until to-day; canal is scaly, with yellow crust. *Calc. sulph.*

Dec. 16. Improved until the 12th, when pain set in; now canal is nearly filled with scales, *débris*, moist and oozing. *Kali phos.*

April 8, 1881. Was relieved entirely by the last treatment, and had no attack until to-day; the canal feels full, and she comes at once for treatment; the same condition as previously; the skin of the canal is removed entire with the crust; the meatus is in better condition after removal than in previous attacks. *Kali phos.*

Jan. 30, 1882. Meatus scaly, congested, slightly sensitive. *Kali phos.*

Dr. M., aged thirty-five years. *Small ulcer on the upper wall of meatus externus.* May 4, 1872. When a child, had ear-ache and offensive discharge. Two years, had three acute attacks of inflammation of the right ear. On examination, the canal is found to be moist, especially the outer third; the inner portion and the *Mt.* being free from pus, but rather dull as regards translucence.

July 8, 1874. The ear has annoyed the gentleman occasionally ever since last date; and, on wiping the canal with cotton, a fetid odor to the discharge is always noticeable. It was impossible to get a view of the roof of the canal until I had a polished steel mirror made for me by Philip H. Schmidt; this introduced through the speculum, the edge of the speculum being brought just within the outer third of the meatus, revealed a small ulcer occupying the space of over five millimetres. The tissues about the ulcer were extremely sensitive to touch, suggestive of localized periostitis. The ulcer was touched with a saturated solution of the *bichromate of potash* on cotton. It caused no suffering, but entire relief of the symptoms. Examination at a later date with the mirror showed only a slight depression at the spot where the ulcer had been.

June, 1883. This gentleman has recently informed me that the one application of bichromate of potash at once and completely cured the whole matter which had annoyed him all his life, and that, too, without the least sign or symptom of local or general disturbance. While I am an enthusiast as regards internal remedies, I should consider myself blameworthy if I failed to pursue a similar course under similar circumstances.

Miss E. B., aged twenty-five years. *Chronic dermatitis.* Dec. 5, 1875. A scrofulous subject; both auricles cracked about the meatus; the meatus moist and scaly; the drumhead covered with accumulation of moist scales. *Conium mac.*

Dec. 11. Better; both ears itch; appearance much the same.

Dec. 17. Intolerable itching in the ears; cracks behind the auricles. *Graphites.*

Dec. 24. The same subjective and objective conditions continue. *Graphites.*

Dec. 29. Acute diffuse inflammation set up in the left ear. *Mercurius vivus.* Diffuse inflammation continued through this month, but under mercurius subsided. In February under *graphites* again.

Feb. 19, 1876. Slight inflammation in the left ear, which yielded to *mercurius*. This has been the history of the case for a number of years.

March 9. Tissues have cleared, and dark wax has formed. *Conium med.*

May 26. Has continued to receive *conium*, and is very much improved.

March 26, 1881. The ears have remained well since 1874. Has a slight catarrhal cold, for which she received *mercurius iod. cum. kali iod.*

Mr. T. S. T., aged twenty-seven years. *Otitis externa circumscripta acuta.* Jan. 14, 1875. For two weeks the right ear has troubled him; has been dull and stopped, without pain, but with subjective noises; right meatus closed with soft cerumen; *Mt.* hidden; H. D. 1-240; two weeks ago he syringed the ear, and brought out some wax. *Fluid cosmoline* was instilled in the canal.

Jan. 18. He had slight pain for two days, and now the ear has been sore for two days; canal injected, and drum-head slightly injected; H. D. 18-240. *Mercurius dulc.*

Jan. 23. Did not experience as much improvement as the removal of the cerumen would have caused if the ear had been free from middle-ear disease; H. D. 38-240; after Politzer, 41-240.

Jan. 25. Has improved; H. D. Politzer, then 48-240; a furuncle in outer third of the canal. *Picric acid.*

Jan. 29. Has improved rapidly; H. D. 16-240.

July 2. Has been well until this date; sensation of soreness in the right meatus, which is partially closed by another furuncle; H. D. 16-20. Prescribed *picric acid.*

July 8. All right.

July 22. All right; H. D. 20-20.

April 14, 1876. The same trouble again; H. D. R. 25-240, L. 5-20; cerumen in both canals, and, when it was removed, R. *Mt.* was found scarred and injected, and the L. *Mt.* slightly injected; mucous membrane of the pharynx red and engorged. *Mercurius dulc.*

April 20. Much better; H. D. R. 9-240, and L. normal.

April 28. Ear seems well; throat slightly sore; H. D. R. and L. 20-20. *Mercurius dulc.*

Miss M. F. M., aged twenty-five years. *Otitis externa diffusa chronica.* Jan. 22, 1876. Suffered four years ago from intermittent-fever, and ten years ago from scarlet-fever. The meatus is very small on both sides; and the chronic diffuse inflammation was such that the smallest speculum was not admitted, and no view of the drum-head could be obtained. The patient had been much worse since the intermittent-fever, for which she received large doses of quinine, and had "salt-rheum." She received *sulphur, conium, arsenicum, silicea, and psorinum* at various times, and was under observation until July 8, 1881, at which time both canals were clear; *Mtt.* were clear, outlines well defined, and not especially thick, or otherwise abnormal. H. D. for the watch was normal. Under psorinum a vesicular eruption formed in and about the auricle and the temporal region, reaching up into the hair over the mastoid process behind. At one time this extended out well on to the cheek. The vesicles ruptured; yellowish crusts formed on the whole surface involved; the crusts exfoliated; and after a month the integument of the auricle and region about the ear became clear, then the right meatus became red. Blepharitis ciliaris set in, which was relieved by *allium cepa* and *rhus.* This was followed by severe occipital head-ache, and extreme sensitiveness to the air, relieved by *silicea* and *cimicifuga.* Since that time, I have had no occasion to prescribe for the lady.

Bertha W., aged twenty-nine. *Nævus of the auricle.* This is congenital, and illustrates the nature of bloody tumors of the auricle. During her youth it was small; but, as she expresses it, "since that I stopped growing, it has grown." It is now the size of a large walnut. This case comes as well under the scope of general surgery, and will be sent to Professor Helmuth's clinic. I will add, in this connection, that, in bloody tumors of the auricle, the indication is to cut off the supply, form a clot, to prevent deformity. In some cases, the tumor has been laid open, and styptic applications used; but a large cicatrix may result, the iron salt being retained. In one case, the galvanic current acted well, stopping the supply, leaving no open ulcer, and but little deformity. Any means that will check the blood-supply to the tumor will cure the condition.[1]

Mrs. G., aged forty. Last April was attacked with erysipelas of the scalp, which extended to the auricle and meatus. Its progress was marked by severe pain, loss of hearing, and at length complete deafness, which lasted over a month: then there was a gradual improvement, but the hearing is not as good as formerly.

This is a case of *otitis externa*, of diffuse character, secondary to erysipelas, extending to the membrana tympani, and undoubtedly involving the middle ear to some extent. You will compare the two ears, and see the uniform capacity of the membrana tympani in the diseased one, in contrast to the normal translucency of the opposite, in which you can trace the outlines of the ossicula. The prognosis is unfavorable: very little change will occur in such a case, but subjective symptoms may be relieved.

Mrs. McCal., aged forty-five. *Otitis externa diffusa.* It is a case of diffuse inflammation, and illustrated what was said concerning the causes of this disease. Two years ago she was attacked similarly, but in a lighter form; and she recovered readily. A few days ago she was attacked by itching in the right ear, followed by soreness, and by signs of inflammation. She presented herself yesterday at the Ophthalmic Hospital Clinic, having relapsed to her former condition from scratching that part till inflammation was renewed. She complained of intense soreness of the meatus, and of pain and itching: the canal was much

1 Professor Helmuth passed needles through the tissues, outside the base of the nævus, and strangulated it: the mass sloughed without secondary flow, and the ulcer healed with very little deformity.

swollen, and the inflammation threatened to extend to the scalp.[1]

This is a typical case of deep-seated inflammation of the external meatus, — diffuse, — involving all the tissues of the canal. I prescribed *calcarea picrata*, and to-day she is very much improved. Had she desisted from scratching, this relapse would probably not have occurred.

In a case like this, hold the patient as long as possible, induce him to undergo treatment for a long time, and thus overcome the dyscrasia by internal medication. To allay the irritation of the auditory canal, and tendency of the patient to constantly pick the ear, no means is better than the application of some petroleum preparation, such as cosmoline or vaseline.

A case has recently been under my care in private practice, — a gentleman afflicted with inflammation of the external ear involving the deeper structure of the skin, as well as the dermoid or epithelial portions. This has decidedly improved under the use of *boracic acid* trit. locally, and *psorinum* internally. This case certainly suggested questions concerning Hahnemann's psora theory, and the importance of destroying the dyscrasia.

I have been surprised by success attained in the similar case of a young lady, also a private patient: therefore I cannot bring her before you, but will detail the case, giving only initials.

L. C., age sixteen. Subject to catarrhal disease for years. Since puberty has suffered with painful menstruation; not robust; has been growing hard of hearing since two years, specially so during the last six months. The hearing for the watch is difficult to settle, because of subjective noises. The voice must be raised, and enunciation be slow; cannot distinguish what is said when a number of voices are heard; has granular pharyngitis; Eustachian tube closed, the left being specially difficult to force by inflation.

[1] The inflammation did involve the auricle and scalp. The patient was critically ill for weeks, and the subsequent treatment was more prolonged than in the former attack.

A gargle of sea-salt was ordered to be used three times a day, and *mercurius dulcis* administered internally. Politzer's method was attempted twice each week. This continued two weeks without manifest improvement. I detected on close examination near the borders of the hair, especially on those parts subject to tension when the hair was dressed, a vesicular eruption. I inquired if she had an eruption in the hair; and, on receiving an affirmative reply, I examined the scalp, and found this same vesicular eruption through the hair, which looked like "the itch" seen on the hands. It was marked by itching and burning. I stopped the prescription directed against tissue change, and gave *psorinum*, a dilution of which I had obtained from that which Dr. Constantine Hering had used in the proving of the virus. At once improvement began, and has continued to the present time, with good indications for the future. The watch now gives 20-20 right ear, 14-20 in the left, with decided gain for the voice.

Whatever may be the opprobrium cast on your treatment by certain wiseacres, if you can get such results, you can bear the obloquy pronounced by those who will not administer a remedy that has an origin not in keeping with their notions. I will not give remedies on empirical or vague indications; but, if a remedy have such a proving as this one, I will use it, "whether derived from purest gold or purest filth," as Dr. Bell says concerning this remedy. When you have many such successes constantly occurring in your practice, it will surely impress you strongly that there is something in the homœopathic law, whatever may be your views as to the theories of miasm, psora, cachexia, or whatever term may be used to explain morbid or curative actions. The success in this showed that there was some cause behind all. I consequently questioned the mother at the first opportunity, if she had had any skin-disease; to which she replied negatively. I pushed my inquiries still farther, and asked whether at school at any time she had contracted an eruption; to which the mother replied that she had contracted an eruption at one time, which showed itself between her fingers. The family physician applied sulphur ointment, or something of the sort, and apparently cured the eruption. Can we see any relation of cause and effect, and psorinum effecting the cure?

There is one thing that the future practitioner must learn, —suppression is not removal; suppression of eruptive symptoms, of periodic symptoms, of painful symptoms, is not removal; the abolition of pain by the use of narcotic remedies is often the destruction of the danger signal, — the loss of the clearest indication to the very remedy that will not only remove the symptom, but cure the disease which causes the symptom.

Cured cases. — Mrs. W. F. L. *Otitis externa parasitica.* Aug. 3, 1872. Three weeks ago had pain in the left ear; no relief since. The pain is a dull aching, pressing in every direction, extending to the tongue, throat, and side of the face. H. D. 12-20; right, -20. Right meatus full of wax; left, the same; removed that in the right without difficulty; removed a portion of that in the left, the mass tearing like shreds of paper, leaving the wall tender and oozing. At another sitting in the evening, removed the rest, leaving the wall of the canal and drum-head denuded, and oozing bloody lymph.

Aug. 4. Pain relieved; the walls of the meatus and drum-head covered with pus, very sensitive on being dried with cotton; hears 24-240.

Aug. 12. The case has been seen each second day until to-day. The question arises, What is this secretion? — the wall throwing off shreds over the entire extent.

Aug. 19. The same condition; shreds not so extensive.

Aug. 26. Less extensive. Shreds will be subjected to microscopic examination.

Sept. 2. Canal and drum-head nearly clear. Examination by various powers, 4-10 to 1-15, gave no satisfactory specimens of aspergillus therein; but from the history and symptoms, as well as from the appearance of the ear, I am sure the case was one of parasitic origin.

Sept. 17, 1877. The last few days the symptoms of August, 1872, have returned. H. D. right, 20-20; left, c-20. The shreds can be turned away from the walls, and towards the middle of the canal. Seizing the mass with a polypus forceps, it was torn away, leaving the entire canal denuded, oozing bloody cerumen. A few shreds remain at the inner third of the meatus. Applied *cosmoline* and *salicylic acid.* The case continued under observation three months.

Oct. 20. It was dismissed free from any sign of parasitic growth. Another examination with the microscope settled the diagnosis of aspergillus. The patient has been seen at intervals since that day, and the trouble has not returned.

Mr. T. F. J. *Aspergillus glaucus.* Oct 14, 1878. Since two months, has been suffering from the pressure of some foreign body in the left ear; thinks it is an accumulation of cerumen; right meatus normal; R. *Mt.* normal; left *Mc.* closed; removed portions of the shreds morning and evening; submitted to microscopic examination, they gave very satisfactory specimens of aspergillus glaucus. The patient continued under treatment during October and November, the portion being removed, and *salicylic acid* used as a parasiticide. *Cosmoline* applied subsequently to overcome any irritation from the acid. There was no reproduction of the growth.

Another interesting case of this disease, in the form of aspergillus nigricans, occurred in the person of a medical practitioner in this city. He had been suffering for some months with irritation of the left ear, and had syringed it a number of times with no relief, but, as he thought, with rather an aggravation of his discomfort. Examination showed in the inner third of the canal, near the drum-head, a dark-colored foreign substance. On touching it with Buck's loop-probe, it was found to be adherent to the floor of the canal; and some degree of force was required to separate it. On examination with a low power, 1-10, it proved to be a very fine specimen of *aspergillus nigricans.* The doctor used a solution of *salicylic acid* in water, with glycerine, preferring that to cosmoline; and at subsequent visits very minute portions were removed by the forcible syringing, after rubbing the seat of the growth with Buck's probe. After repeated trials of the aqueous solution, an alcoholic solution gave complete relief, followed with cosmoline to relieve any irritation caused by the acid.

Mr. E. P. H. Nov. 27, 1872. The history of this case was similar to the preceding ones, save that the parasite found a favorable condition for development in ulceration of the canal, which persisted for a long time after the growth was destroyed by a one per cent solution of carbolic acid.

The ulceration continued for nearly a year after the destruction of the growth, although the hearing became 18-20.

In a case seen at the Ophthalmic Hospital, I obtained the largest mass of fungus that I have ever seen, or removed entire from the canal. It was a specimen of aspergillus nigricans, was mounted for me by Professor T. F. Allen, M.D., and given to Professor J. W. S. Arnold, M.D, for the purpose of having it photographed. Immediately afterwards I was prostrated by severe illness, and for nearly a year was unable to pursue the matter. Upon inquiry, I found that the specimen was taken to Professor Arnold's rooms at the University Medical College; and to his disappointment, as well as my own, the specimen could not be found.

G. H. B. *Effects of explosion.* Nov. 1, 1880. Two weeks ago, a gun exploded near the left meatus, causing slight pain, and a hissing noise set in. The meatus is swollen, the drum-head sodden, the throat red, catarrhal; H. D. c-20. *Belladonna* during the attacks of pain; *hepar sulph.* every three hours at other times.

Nov. 3. Very much improved, meatus swollen, tissues dry.

Nov. 5. Had severe pain last night; three furuncles in left meatus; no view of the drum-head; continued *belladonna* with *picric acid;* much better; furuncles have ruptured, and are oozing.

Nov. 10. Soreness of the ear rather than pain; meatus sensitive, swollen. Continue *picric acid.*

Nov. 12. Hears 12-240; meatus closed by scales.

Nov. 16. Much improved; hears 19-240; canal clearer; drum-head thick. *Kali mur.*

Nov. 22. The same.

Nov. 26. Not much better; 27-240. *Kali sulph.*

Nov. 29. Has taken severe cold; hearing not so well; canal scaly; drum-head not much clearer; Eustachian tube dilatable. Continue *kali sulph.*

Dec. 11. Better; hears 38-240. Continue *kali sulph.*

Dec. 22. Better; hears 42-240.

Jan. 7. About the same; hears 42-240.

Feb. 7. About the same appearance of the canal, and drum-head about the same.

In this case, I am satisfied that the explosion of the gun was only the last element in a process which had been of long standing, namely, catarrhal disease of the middle ear, and that the acute condition set up was simply superadded to a chronic condition already existing. The patient made only a partial recovery.

Master George B. F., age eight years. *Foreign body.* Sept. 5, 1873. One week ago, while at play, put a cherry-stone in his right meatus: efforts had been made to remove it, with no success. The meatus was torn, and any attempt at examination was resisted with all the power the lad could bring to bear. He was etherized; and, on wiping the canal, the round surface of the stone presented, filling the entire space. A probe bent at right angles was passed between the stone and the posterior wall of the canal, till it could be turned, beyond the stone. Steady traction brought the stone without injury to the walls. *Mt.* perforated and ulcerated. *Arnica.*

Sept. 10. *Me.* pus, *Mt.* perf., ulceration the same; Et. dilatable; tendency to mastoid inflammation. *Capsicum.*

Sept. 18. Better; very little discharge of pus; perforation has clearly defined edges. *Silicea.*

Sept. 25. No discharge for two days; mucous secretion in place of pus. *Silicea.*

Oct. 1. Improving; discharge slight; patient gained slowly till Oct. 22, when *capsicum* was given.

Nov. 19. Discharge ceased; H. D. 3-20; perforation small.

The father of the lad declined further treatment. I regret not being able to report the final result.

Diseases of the External Ear. — Summary of Remedies.

Aconite. — Anxiety; restless tossing; sensitive to light and noise; apprehensive; fears death; fulness and heaviness of head and brain; burning, congestive headache; face bloated, red and hot, but pales on rising; burning thirst. Indicated usually in the early stages of disease, or when temperature rises in later history.

Arsenicum album. — In otitis externa diffusa. Objective symptoms: Tissues of meatus red, infiltrated, oozing clear watery fluid, in some cases vesicles, in others the tissues thin, dry, and scurfy. Subjective symptoms: Burning and itching; itching is aggravated by scratching, and ameliorated by heat.

The iodide of arsenic has proved more effective in some instances than arsenicum album.

Antimonium crudum. — In cases without objective symptoms, the subjective ones being heat and tension, aggravated by heat. Otitis media catarrhalis chronica, left; with the symptoms due to lesion of the middle ear. The heat of the auricle and meatus was increased by sunlight; heat from the stove, wrapping, or even turning upon that side in sleep, would increase the heat, and waken the patient. Antimonium crudum caused transfer of the heat to the scalp. Glonoin has since controlled the latter condition.

Belladonna. — Beating headache; throbbing in brain, with sensation as if it were loose in forehead; worse from walking or rising; head sore to touch; tearing pains in ear; mouth dry and hot; posterior wall of pharynx dry and glazed; bright red. The fever is marked by alternate chill and heat, or internal chill with external burning. The patient starts suddenly, rather than constant motion. (Compare aconite, ferrum, and gelsemium.)

Calcarea carbonica.—One of the most valuable remedies in suppurative inflammation of the external ear, in scrofulous subjects. The meatus filled with cheesy pus, the derma thickened and red. Often the dermoid layer of the membrana tympani destroyed by ulceration, and covered with exuberant granulations, which may fill the meatus, the structure of these polypi being of the simple cellular class. There is a lack of subjective symptoms: in some a pulsation is noticed.

Calcarea picrata.—Indicated by clinical experience for perifollicular inflammation. The extreme prostration of picric acid is relieved by this salt, also.

Carbo vegetabilis is indicated in otitis externa diffusa chronica; objective symptoms being a dry furfuraceous eruption, a pityriasis, thin, dry epithelial scales thrown off, unattended with marked inflammatory signs. In some cases the detritus is moist, yellow, and fetid. Subjective symptoms: Itching and heat deep in the ear as well as in the meatus, causing an inclination to swallow. The ears feel stopped. These symptoms suggest the relation of the meatus to the tympanum and Eustachian tube. Carbo veg. is valuable in granular pharyngitis with the above symptoms, the expectoration being small masses of mucus, easily raised. Under the use of carbo veg. the secretion of wax is reestablished in many cases.

Chamomilla is of value when the patient is extremely intolerant of pain. Specially valuable in diseases of children and sensitive female adults.

Conium maculatum.—Valuable in hypersecretion from the ceruminous glands. Objective symptoms: Accumulation of soft cerumen, of normal color, and clinically, that which resembles mouldy paper, and is mixed with pus. The subjective symptoms are a sense of fulness, roaring, and humming; diminution of hearing, which is ameliorated, in many cases, by pulling the auricle, as this straightens the canal, and makes a passage for the sound-waves beside the mass of cerumen.

Ferrum phosphoricum. — From Schüssler's "tissue remedies:" to be given in the early stages of inflammation, or later when temperature rises, and pulse increases. One guiding symptom observed clinically is the noticeable pulsation in the ears: every impulse of the heart is felt there. Compare aconite.

Graphites. — The characteristic symptoms of this remedy, as regards the integuments and nails, are noticeable in this locality. The objective symptoms are dryness and cracking of the tissues of the meatus and auricle, particularly behind the latter; deep fissures in many cases. The pus, both in the meatus and about the auricle, is usually thick, and forms crusts very rapidly. The subjective symptoms are itching and soreness, not to the degree that may be called sensitive (see hepar).

Hepar sulphurus calcarea. — The appearances that suggest this remedy are those of an indolent ulcer, corroding, and very sensitive to touch; the pus fetid and thin, or, if the membrana tympani be perforated, mingled with mucus. The subjective conditions are itching in the meatus, with soreness on attempting to bore with the finger. Better by wrapping.

Kali muriaticum (Schüssler). — Chronic dermatitis; moist, excessive exfoliation of epithelial layer; in ulceration, pus whitish; granular conditions of inner third of meatus and on *Mt.*

Kali phosphoricum (Schüssler). — Atrophic conditions in old people, tissues dry, scaly; lack of vitality.

Kali sulphuricum (Schüssler). — Conditions similar to the muriate: the bright yellow color of the pus is a guiding indication.

Mercurius vivus. — Although this remedy is specially indicated in diseases of the middle ear, yet it may be used as an intercurrent remedy in disease of the meatus, with conium and carbo vegetabilis. In January, 1873, we treated Mrs. T., aged forty-eight, a thin, small woman, dark hair and eyes. Had discharge from R. since childhood. One month later took cold, and lost the power in the L. Hears voice 20-80; watch, 2-240 R. and L. *Mc. ext.* scales and shreds; R. *Mt.* irregular, retracted, but movable; Et. dilatable; throat granular. The outermost dermoid layer of the membrana tympani was repeatedly thrown off during the course of treatment, the meatus filling with cerumen and scales. Carbo veg. and conium made remarkable change in one year, the hearing for voice becoming nearly normal; watch, R. 10-240; L. 10-240. The second year she was seen at intervals of about two months. During the third year, on one occasion, some of the shreds were accidentally carried very near the terminal filaments of the olfactory nerve, and the decided *coppery* odor was noticed. Merc. was given with most marked relief to the local as well as general condition of the patient.

Mezereum.—This remedy is very effective in relieving intense itching in the auditory meatus. The sensation extends to the Eustachian tube. (Compare causticum and nux vomica.) One guiding symptom is the sensation "as if the cold air reached the tympanum."

Picric acid.—This remedy is indicated by the recent proving in furuncular or circumscribed inflammation of the meatus; yet, in the chronic or subacute forms, it has delighted patients and surgeons. In debilitated cases, with redness and localized tenderness of the meatus, it acts like magic.

Plantago major.—For local use, to mitigate the intense pain of circumscribed or diffuse inflammation. (Compare middle ear.)

Psorinum.—External ears, raw, red oozing; scabs form; sore pain behind ears. Otorrhœa very offensive, purulent, watery. Pustules on and behind concha. Scabby eczema behind right ear. Herpes from ears to cheeks. Scurfs form, and scale off; yellow discharge from under scurfs; itching intolerable. This remedy has been repeatedly confirmed, both in clinical and private practice.

Pulsatilla. — Specially indicated in circumscribed inflammation in sensitive women, with extreme chilliness; darting pains, worse in the evening; better by exposure to cool air. (Compare hepar.)

Silicea corresponds to an ulceration of the tissues of the membrana tympani and inner extremity of the meatus, which is deep, slow to heal, covered with thick yellow pus: the pus tends to the formation of a firm scab, which adheres closely to the ulcer, and, on removal, reveals the above condition. The tenderness is marked on touching with probe and cotton, but not so extreme as hepar sulph. calc. One subjective symptom that suggests this remedy is a hissing sound accompanying the purulent discharge.

Sulphur. — This remedy applies to many and varied, sometimes opposite, conditions. The appearance of ulceration, with perforation of the *Mt.*, is usually thick edges, with thin pus, fetid, tending to crusts. The ulcers show no disposition to heal. Subjective symptoms are burning and itching, or pricking; a sensation of tension or drawing deep in the meatus, or a pulsation in the same, would suggest the remedy; but these objective and subjective symptoms are usually most marked in diseases of the middle ear.

Tellurium is of great value in acute inflammation of auricle, meatus, and external surface of membrana tympani, — a condition similar to phlyctænular conjunctivitis; yet the remedy has proved curative in chronic suppurative inflammation, with the characteristic discharge, watery, excoriating, and very fetid, smelling like fish-pickle. The condition of tissues in the prover, Dr. Carroll Dunham, indicates that it should cure suppuration of the middle ear as well.

Part Second.

DISEASES OF THE MIDDLE EAR.

FOURTH LECTURE.

ACUTE CATARRHAL INFLAMMATION OF THE MIDDLE EAR.

Gentlemen, — The nomenclature of diseases of the middle ear is a matter of importance; for, if it be true that language is necessary to thought, a clear and definite classification will help us to avoid erroneous ideas of disease.

I shall follow the classification suggested by Professor Roosa: First, acute catarrhal inflammation; second, subacute catarrhal inflammation; third, chronic non-suppurative inflammation, in the two forms, catarrhal and proliferous; fourth, acute suppurative inflammation; fifth, chronic suppurative inflammation; sixth, consequences of chronic suppurative inflammation.

Catarrhal inflammation is not a local, but a constitutional, disease, and requires constitutional treatment. It is impossible to deal successfully with these catarrhal diseases, unless we attend to the detail of hygiene. At times we may use remedies alone, and meet with good results; while, again, we may fail because of neglect in the direction of hygiene. There is a tendency, on the part of many practitioners, to follow a routine in all cases that come to them, never leaving a certain fixed mode of practice. It is to this routine mode of treatment that so many failures in the cure of disease are chargeable, whether it be directed towards hygiene alone, excluding medicine, or the use of medicine, excluding the essentials of personal and general hygiene. To impress more fully the importance of careful investigation in every instance, let me refer to a case: —

A lady came to me who had for two years been suffering from what was said by a number of physicians to be catarrh. It was catarrh treated in the routine way, "powders, powders, powders." The patient had never been subjected to an examination of the naso-pharyngeal tract. On investigation, I found the nares filled with gelatinous masses which perpetuated the condition. It was necessary to use local, mechanical, and constitutional means to bring about a cure.

Statistics should show that *Acute Catarrhal Inflammation of the Middle Ear* is a very frequent form of disease, though it is not so reported. Many a serious result, affecting the hearing of the adult, may be traced to what may be considered a trivial affair in the child, "only an ear-ache."

Symptoms. — The symptoms of acute catarrhal inflammation are pain, fulness, noises. Objective symptoms are injection of the membrane, bulging of the same, impairment of hearing, naso-pharyngeal catarrh, general fever. The pain is deep-seated, usually severe at night, mitigated during the day, and often leading the patient to suppose that his trouble has passed: it may pass, but only to return with the approaching night, and, unless treated, grows more and more severe on successive nights, until it passes into the exudative form of disease, or on to the suppurative form. The sense of fulness may precede the pain for days, and is often associated with itching. This pressure increases from day to day, until the pain ensues. Often, before the pain, subjective noises will be recognized; but the impairment of hearing is not usual until after the pain has set in. On inspection, the membrana tympani will be found injected, especially about the handle of the malleus, Shrapnell's membrane, or its entire periphery. This injection may pass away, and resolution occur without more serious symptoms. Later the injection amounts to absolute congestion of the entire membrane; and, as the accumulation increases in the cavity of the tympanum, the inward concavity of the drumhead will be changed to a bulging into the canal. This may be excessive, and if relieved by paracentesis, and removal of the accumulated secretions of the tympanum, will be entirely overcome as the case progresses to resolution.

Diagnosis. — The impairment of hearing is very great in many mild cases, and is a strong diagnostic point between otitis and otalgia neuralgica. In all cases, there is more or less naso-pharyngeal catarrh, this being the remote cause of the middle-ear disease. Febrile disturbance is another symptom which will enable you to decide between otitis and otalgia. The rapid pulse and high temperature are not found with otalgia.

Causes. — The causes of acute inflammation of the middle ear are those of catarrh in general; and, when we enter upon this field of investigation, we find it very wide. Climatic conditions, personal habits, clothing, bathing, eating and drinking, all are involved in this discussion. It is noticeable, that on the Atlantic belt, on the river-courses, and in the lake regions, there is a general complaint that the excessive moisture interferes with the function of the skin, to such a degree that the mucous membranes of the body have to do double duty: such being the case, only the most rigid regimen as to personal habits, clothing, and food, will so sustain the functions of the external surface of the body that internal organs will not be burdened. Space will not suffice to consider in all its bearings the matter of climate; but, in general, a dry, uniform atmosphere will overcome the disease, even in serious cases, when a change of location is feasible: if not, all means which will preserve the function of the skin must be carefully employed. Bathing is a matter which has been abused, yet it is one of the most important of personal habits. Excessive bathing frustrates the very object in view. Persons are exhausted by excessive bathing, as they lack power of re-action; and hence their vitality is drawn upon to a degree which in some cases has been fatal. A warm bath once a week is sufficient for personal cleanliness; but a tepid bath at night if the subject has a low degree of vitality, or a cool bath in the morning if the patient be otherwise vigorous, with hand-rubbing, or rubbing with a coarse towel over the entire body, until re-action ensues, is invaluable. The clothing should be such as to allow ready elimination of the insensible perspiration, and yet of such material as to prevent radiation of the animal heat. Special attention should be given to the clothing of the extremities and feet. Many a catarrhal patient, especially among women and children, is wrapped and bundled about the trunk of the body, while the extremities suffer from exposure.

I am satisfied that one's habits, in the matter of food and drink, have much to do with the excessive accumulation of mucus. While I am not an advocate of an exclusively vegetable diet, I am satisfied by personal experience, as by observation, that excess of animal diet, especially when of a stimulating nature, and associated with stimulating drinks and the use of narcotics, is a most potent factor in the supply of material which must undergo destructive metamorphosis; as it can serve no part in the maintenance or upbuilding of the tissues of the human body. Diseases of a catarrhal nature which affect the pharynx, must necessarily involve the Eustachian tube, and are the direct cause of middle-ear diseases. Among the most noticeable direct causes of acute inflammation of the middle ear are the use of the nasal douche, and snuffing water.

Prognosis. — Unfavorable, if under adverse climatic conditions; favorable, with the best conditions, and with such treatment as later years have demonstrated to be practicable.

Treatment, medicinal. — *Belladonna*, *chamomilla*, *gelsemium*, *hepar sulph. calcarea*, *mercurius*, and *pulsatilla* will be indicated in acute catarrhal inflammation, — the first three and the last for the mitigation of pain, the other two against the destructive changes of tissue.

Belladonna is indicated when the pain is shooting, beating, darting, associated with marked chilliness, not especially marked thirst, and great restlessness.

Chamomilla is more particularly indicated in children, or when the patients are very intolerant of pain, being unable to control themselves under what seems rather trivial suffering.

Gelsemium is especially for the more apathetic, quiet condition, fever not marked, the patient inclined to stupor, the head-symptoms being those of pressure and tension, relieved by outward pressure or binding.

Pulsatilla is indicated in women and children, persons of an extremely sensitive nature, and in those cases where there is relief of most symptoms from being in the open air.

Hepar sulph. calcarea will often ward off suppuration when the patient is extremely sensitive to air, and is relieved by wrapping, and by warm, dry applications. This is in contrast to *mercurius,* under which remedy the patient suffers with excessive perspirations of a sticky or greasy nature, which do not relieve the general suffering, but annoy the patient. The perspiration is similar to that of mercurius cases, when hepar relieves, but not so general or continuous. There is often, with the hepar patient, a localized tenderness of the tissues about the ear, especially in front; whereas, in mercurius, the tenderness is more general about the ear, and extends down the neck beside the jaw. Careful study of symptomology will often suggest some other remedies, but these are more usually indicated.

Instrumental means of treatment. — Warm applications are usually very grateful, — dry heat in the form of the salt-bag, or warm vapor, warm water dropped into the meatus from a sponge, drop by drop, as warm as can be tolerated. This is made more effective by the addition of a few drops of *aconite, belladonna,* or *plantago.* Some writers have advised the use of warm applications in the form of poultices, or fomentations of herbs, especially of hops This is reprehensible practice: suppuration is often induced by such applications. Some of the worst cases of suppurative disease, involving the mastoid, have resulted from the abuse of warm, moist applications.

Paracentesis of the membrana tympani has been advised, followed by inflation, and removal of the mucus. I believe, however, that the use of the indicated remedies, with inflation by Politzer's method as soon as practicable, will, in the vast majority of cases, cut short the attack without paracentesis, and that the mucous accumulations will be absorbed. Later treatment, as prophylactic, must bear upon the naso-pharyngeal condition.

Subacute catarrhal inflammation is, strictly, either the antecedent of the acute form, or the resultant of an acute inflammation, standing between resolution on the one hand, and chronic catarrhal inflammation on the other.

FIFTH LECTURE.

CHRONIC CATARRHAL INFLAMMATION OF THE MIDDLE EAR.

Gentlemen, the next subject in the order of our classification is *Chronic Catarrhal Inflammation of the Middle Ear.*

History. — It is necessarily the successor of repeated attacks of acute catarrhal inflammation, or of subacute catarrhal inflammation if the patient should be fortunate enough to escape acute suffering; and yet we can hardly say, "fortunate enough," because acute symptoms often lead to a careful investigation of one's condition, and the recognition of a subacute or chronic phase of the disease. I desire to impress upon your minds the very great importance of an early recognition of this condition. The fact that it is gradual, and insidious in its approach, is the serious feature of its history. The degree of hearing that one may lose, and yet be unaware of the failure of the function, is very remarkable. In this respect, disease of the middle ear stands in marked contrast to disease of the eye.

The effect of cold upon the eye in causing congestion is at once recognized, even in the less grave affections; as conjunctivitis, superficial keratitis, etc., receives prompt attention, and that attention continues until a cure is reached. Similar congestion of the middle ear may occur again and again, and be neglected until tissue changes take place, which it is impossible to overcome in the later stages of the disease. The reasons why this is impossible, I shall endeavor to emphasize by calling your attention to these charts of the middle ear, showing you the relation between the membrana tympani, the tympanic walls, and the ossicula themselves. Let me assist you by calling your attention to the diagram which I will sketch upon the board.

You will notice the plane of the membrana tympani in a normal condition, as shown by the cross-section cut perpendicularly through the middle of the tympanum. This divides the ossicula nearly in the middle; and you will understand, that, if any cause acts to close the passage of air by way of the Eustachian tube, there is no atmospheric pressure to counterbalance the pressure exerted upon the drum-head by the external air. Thus, the drum-head is forced inward, the handle of the malleus approximating to the incus and to the tympanic wall. Now, if to this pressure, as an element of the change of relation of these parts, there be added changes in the structure of the membrane itself, you will see, that, by relaxation, it may be actually in apposition with the thickened mucous membrane of the tympanic wall. When this occurs, adhesions form, similar to adhesion between the pleura pulmonalis and pleura costalis in pleuritis. When these adhesions are established, they remain through life. Various operations have been suggested, as we shall see when speaking of instrumental treatment; but they have proved futile in most instances. One of the earliest, most persistent, most annoying results of this adhesive process is the production of subjective noises. If you will close your eyes while sitting here, and repeatedly strike or press your fingers upon the closed eyelids, you will understand by analogy what takes place with every vibration caused by similar pressure, if the stapes be forced upon the labyrinthine fluids. Now, this subjective condition is one of the earliest intimations in the history of chronic catarrhal disease; and I desire to enforce this point, that you may not neglect it when you have to deal with it in practice. It is true that this condition may occur when not associated with chronic catarrhal disease, as in prostrating diseases, associated with great loss of blood, anæmia, or vertigo and faintness, secondary to some organic disease. But a careful study of these cases will reveal

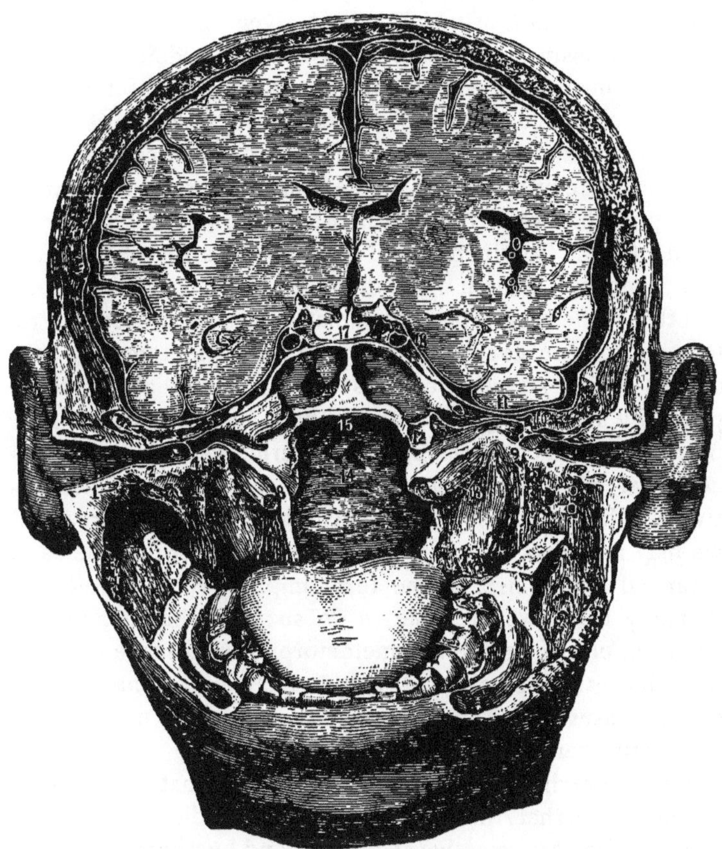

SECTION OF THE HEAD, SHOWING THE DIVISIONS OF THE EAR AND THE NASO PHARYNGEAL CAVITY. (AFTER A PHOTOGRAPH. RUDINGER.)

1. Cartilage of external auditory canal. 2. Osseous auditory canal. 3, 4. Membranæ tympanorum. 5. Cavity of the tympanum. 6. Dilator muscle of the Eustachian tube. 7. Levator palati muscle. 8. Mucous membrane of the pharyngeal orifice of the tube. 9. Left membrana tympani. 10. Handle of the malleus and short process. 11. Tensor tympani muscle. 12. Mucous membrane of the membranous portion of the tube, perforated by a needle. 13. Levator veli palati muscle. 14. Mucous membrane of the posterior surface of the pharynx. 15. Mucous membrane of the pharynx, attached to the lower surface of the body of the sphenoid bone. 16. Sphenoidal sinus. 17. Hypophysis cerebri, and its relations to the cerebral arteries and the cavernous sinus. (From Roosa's Treatise.)

sufficient causes for the subjective symptoms; whereas, in chronic catarrhal inflammation, you will not usually find other explanation than that which is here given.

In the classification of the disease of the middle ear, which has previously existed, we find a great variety of terms expressive of the condition: for instance, sclerosis, otitis media hypertrophica, otitis media hyperplastica, tubal catarrh, tympanal catarrh, proliferation. The first is a correct translation of expressions used by Continental writers; the second and third, of some Continental and English aurists; tubal catarrh and tympanal catarrh have been used by English writers mostly; while proliferous inflammation is a term used by Professor Roosa to cover the whole field of symptoms and conditions which stand in contrast to the catarrhal condition, primarily and simply. Grouping these together, I have here placed, in contrast to them, the two terms, catarrhal and post-catarrhal; as the term *post-catarrhal* is, to my mind, descriptive of a large majority of these cases. While it is true that we may not always be able to trace the history of the early catarrhal experience, as the term implies, yet it is true that in many cases we are able to do so; and the later history has been one of retrograde metamorphosis, absence of secretion, dryness, sclerosis, and, subjectively, of all the features of the cases which were previously grouped under those descriptive terms.

By "catarrhal," then, we understand that division of chronic catarrhal inflammation of the middle ear which is characterized by excessive secretion; and by "post-catarrhal," the group of cases which stand in marked contrast to the former. The catarrhal form corresponds to what some writers have called *humida*, and the post-catarrhal to that which some writers have called *sicca*.

Subjective symptoms of the catarrhal form. — Secretion in the posterior nares, pharynx, Eustachian tube, and tympanum. Its progress is not insidious. It is characterized by fulness of all the naso-pharyngeal tract. Sounds in the ear, not necessarily excessive. People are said to speak low.

Subjective symptoms of the post-catarrhal form. — Very little secretion in the naso-pharyngeal tract; the loss of hearing being the earliest, and sometimes the only fact, which the patient has noticed. The naso-pharyngeal tract characterized by thinness of the tissue; the auditory canal the same, dry, and enlarged in calibre. The subjective noise is excessive, and very distressing; and you will often find associated with this, the peculiar symptom of the person hearing better in a noise. The explanation of this symptom has been attempted by various authors. The one which is most satisfactory to my own mind is, that the greater vibrations caused by loud noises, as the sounds of machinery, the rattle of cars, etc., bring into vibration the entire auditory mechanism, and, while it is then in functional activity, the lesser vibrations are conveyed and recognized; which is not the case when the patient is in a perfectly quiet room.

Objective symptoms of the catarrhal form. — Deafness; also objective changes in the membrana tympani, as thickening and opacity. Changes in the tympanum, by which its walls are increased in thickness, and the calibre necessarily diminished: the Eustachian tube is similarly affected. The same is noticed in the naso-pharyngeal tract. The diameter of the auditory canal is usually normal.

Objective symptoms of the post-catarrhal form. — The same contrasts are noticed, which you will more clearly understand when we consider the pathological features.

Pathological changes, catarrhal form. — First, thickened mucous membrane; second, excessive mucous secretions; third, lymphy secretions.

Pathological changes, post-catarrhal form. — Connective tissue changes, binding the mucous membrane in various directions. The bony walls hypertrophied, Eustachian tube obstructed by bands; stapes anchylosed, exostoses, incus and malleus anchylosed, mastoid cells closed by bands, tendon of the tensor tympani muscle adherent to the ossicula, or to the tympanic wall; atrophic degeneration of the tensor tympani muscle. While it has been shown, that, even in the fœtus, there are similar formations, yet it is also shown that these are absorbed in the early life of the child; and it is begging the question to assert that the formations in the post-catarrhal form are coincident with fœtal life, for any careful observer will admit that he has seen these changes grow more and more marked in cases which I have called post-catarrhal.

Causes. — The same causes which are effective in acute catarrhal disease are also potent here, and should receive the same attention. Phthisis, syphilis, scrofula, in fact any cachexia tending to degeneration of tissue, to metamorphosis either in the direction of excess or deficiency, will lie at the foundation of these two forms of disease. The remote causes are those which lie far back, in the acute or subacute catarrhal attacks. There is a certain class of cases which have caused considerable discussion, in which the sudden failure of function would seem to be due to a sudden loss of nervous force. Pregnancy has seemed to be the exciting cause in not a few of these cases; each parturition being marked by a greater loss of function, — a loss which is seldom repaired to any extent. In some cases, I have been unable to attribute any cause for the sudden and irreparable partial loss of function. Any means which would act temporarily to increase the tone of the nervous energy of the individual, would cause corresponding temporary improvement of the function.

Treatment. — Instrumental and medicinal. The treatment, as regards both instrumental and medicinal methods, will be divided according as the case is one of catarrhal, or so-called post-catarrhal, nature. For the catarrhal strictly, much may be done; for the post-catarrhal, — save in a few instances, where the muscular tissues can be restored to partial function, — very little, from the nature of the case, has been, or can be, accomplished. First, consider the purely catarrhal cases. The treatment should be directed to overcome the excessive secretion, and to the reduction of the thickened mucous membrane, thereby increasing the calibre of the tympanum and the Eustachian tube; and, to be in keeping with our views of treatment, this must be done mainly by internal medication. You will understand, therefore, why I consider as worse than useless, persistent interference with the mucous membrane of the pharynx and Eustachian tube, by forcible means of dilatation, and acrid or caustic applications to the same region. It has occurred that the continuous exhibition of the clearly indicated remedy has so reduced the thickened mucous membrane, and restored the patency of the Eustachian tube, that the function of the same has been suddenly restored, without instrumental interference. The patient has experienced the subjective symptom of a loud explosion or detonation in the head, and, to his delight and astonishment, has found his function measurably restored. If much can be done, and is done, by our friends of the opposite side when using mechanical treatment simply and purely, much more can be done by us when instrumental treatment is supplemented by indicated remedies. The use of the various instruments for inflation will be shown in connection with cases. I therefore pass to a description of various operative procedures, which have been suggested, and show why they have almost universally failed to be of value.

SIXTH LECTURE.

CHRONIC CATARRHAL INFLAMMATION OF THE MIDDLE EAR CONTINUED.

Gentlemen, — Great stress is justly laid by some writers upon the constitutional treatment as well as the hygienic care of the patient ; but beyond the administration of forms of potash and lime, in massive doses, there is little to be obtained from the so-called "regular" literature which will be helpful to us. One conclusion which the best authors reach, is that applications of a caustic or escharotic nature are admissible only in the catarrhal form. What I have said in regard to the use of the douche as an instrument of treatment, holds good in the chronic catarrhal form. While it is true that there is a temporary relief of the naso-pharyngeal symptoms, the method is a great source of danger. The same is true of Gruber's method, which consists of injecting various medicated fluids by using a small-bulb syringe ; introducing the tip into the anterior nares, and forcing the contents through into the pharynx. When it is desired to flood the tympanum, the patient's head is turned toward the side which it is desired to flood ; and he is instructed to hold the nose, and blow, while the head is down. Bear in mind that this should never be done without the full recognition of the dangers involved, and the necessity that the patient should be kept under the surgeon's eye, that he may guard him against any acute disease, — by the use of inflation to dissipate the contents of the tympanum into the mastoid cells, or outward again by the way of the Eustachian tube.

These passages and cavities are, by nature's arrangement, for air only; and air or vapor is better adapted as a medium of medication.

Bougies are inadmissible. The fact that the Eustachian tube, in its inner extremity, passes through bony walls, increases the difficulty of the passage of a bougie, and thereby makes it a less feasible instrument than in the treatment of other narrow passages of the body. In connection with the cases to which I have already called your attention, the use of electricity was mentioned as a means of stimulating the degenerated muscles of the tympanic cavity, and I urge you to qualify yourselves to properly apply this useful and yet dangerous agent: you will obtain gratifying results in those cases which have the factor of degenerated muscular tissue as one of their prominent lesions.

The question may arise, How long, or when, is it advisable to treat cases? Persons suffering from this form of disease must understand that they will require every year more or less attention. It will be impossible to gain any degree of audition, or even to retain the remaining degree, without attention some portion of the year. During the cold season, repeated exposures will cause an accession of serious symptoms. These must be watched and treated. During the time of the constitutional change, which a person undergoes in the spring or early summer months, I am satisfied that we can render these patients more service than at any other time of the year. During the summer they will need very little care, and usually will not retrograde until the late fall or early winter months. Therefore, by careful attention, you will be enabled to protect from complete loss of function, and possibly add each year a little power to that existing when first seen.

The next item of treatment we have to notice is that of the various operations upon the membrana tympani. So far back as 1650, the question was raised, whether incision of the membrane would be a justifiable and useful operation. As early as 1760 the operation was made by one Eli, whom

Roosa reports as probably a charlatan. To Sir Astley Cooper is due the introduction of the operation, and demonstration of its advantages. It fell, however, into disrepute, but was revived by Schwartze for acute troubles, removal of mucous accumulations, etc.

Politzer, in 1845, suggested the introduction of the eyelet into the membrane; because it was found by Schwartze and others next to impossible to maintain a permanent opening. Voltolini succeeded in establishing an opening for a greater length of time than others had been able to accomplish, by the use of the galvano-cautery puncture. He found that the use of this instrument was followed by less hemorrhage, by cicatricial tissue that did not close the aperture as readily as when other methods were used. Politzer's eyelet was used as a means of retaining a permanent opening. The eyelet is simply a small one of ivory or bone, which is placed in a slit-like incision at some selected point of the membrana tympani, and is allowed to remain until the tissue heals about it, in a somewhat similar manner as it does about the gastric canula for experimental investigation of the function of the stomach. In a number of cases, however, suppurative inflammation has set in; and the eyelet in one case dropped within the tympanum, and was the source of serious inflammation before it was finally removed. The results obtained by these efforts at permanent openings have been neither uniform nor satisfactory. In a number of cases which have come under my observation, in which the galvano-cautery was used, the patients' statements were to the effect that the tinnitus was greater after the operation, while the hearing was not permanently improved. Careful examination of the patients who had been under my care, and who were subsequently operated upon in this manner, showed that the cicatricial tissue resulting from the galvanic puncture formed more extensive adhesions than previously existed.

Gruber, in 1863, introduced to the profession his operation called myringodectomy. A triangular flap was cut out of the membrana tympani, with a hope that a permanent opening would be maintained by the healing of the edges of the wound in such a way that it would not be completely closed. The operation is pronounced dangerous, both by reason of excessive hemorrhage, and by the subsequent suppuration which almost always followed; and, in those cases in which both of these dangers have been avoided, the operation is found to bear no proportionately favorable results to iridectomy. Weber Liel suggested a division of the tendon of the tensor tympani muscle. This operation is based on good physiological reasons; for the release of the tendon of the tensor tympani muscle should overcome the inward pressure upon the ossicula, thereby overcoming the pressure upon the labyrinthine fluids. The results of this operation, again, are not uniformly successful. Division of the posterior folds is suggested by Politzer. This is open to the same objections as hold regarding cicatricial tissue already mentioned. Prout has operated for the division of the adhesions which exist between the membrana tympani and the ossicula, or the tympanic walls. This seems the most feasible of all operations, and is the only one which my experience warrants. The difficulties encountered in locating adhesions, and certainly in determining their extent, are appreciated by one who has had any extended experience in the use of Siegel's otoscope, as used for the determination of this fact.

Hinton's operation of incision for the removal of accumulations of mucus or lymph, is a practical suggestion, and has been generally accepted and practised, whereas all the others have fallen into disrepute. Some years ago it occurred to me, while using Siegel's otoscope for diagnosis, that it might be applied as a means of treatment; and I attached the mouthpiece of the instrument to an exhausting-pump, and have since used it more or less. About a year after my first use of it, Dr. Howard Pinckney published an article suggesting a similar use of Siegel's otoscope, using a common stomach-pump as a means of exhausting the air. This treatment is often productive of the cessation of subjective sounds, as well as of the sensation of stuffiness or fulness in the ear.

Medicinal treatment. — The medicinal agents used for this form of disease of middle ear divide into two classes, as the drugs have excess of secretion, or are marked by dryness of the mucous membrane in their physiological action. Prominent among the remedies for the first form characterized by thickening, hyperæmia, hyperplasma of the mucous membrane, are *baryta muriatica, calcarea iodata, calcarea phosphorica, conium, gelsemium, hepar sulph., hydrastis, iodine, kali muriaticum, kali hydriodicum, mercurius, phytolacca, pulsatilla, sanguinaria nitrate, teucrium.* For the opposite state of atrophy, *carbo veg., causticum, cinchona, graphites, iodine, kali phosphor., kali hydriodicum, magnesia phosphorica, petroleum, phosphorus, silicea.*

SEVENTH LECTURE.

ACUTE SUPPURATIVE INFLAMMATION OF THE MIDDLE EAR.

Gentlemen, — We will consider to-day the subject of *Acute Suppuration.* It is to be remarked, that the acute catarrh thickens the tissue, whereas acute suppuration destroys the tissue; i.e., that repeated attacks of acute catarrhal inflammation, unless passing over into chronic catarrhal inflammation, causes a proliferation and thickening of the tissues of the tympanum. Acute suppurative inflammation, as already marked, tends to the destruction and perforation of the membrana tympani. There are apparent exceptions to this in certain cases, where there is no pain, no appearance of moisture, until suddenly pus is found flowing from the ear. In the great majority of cases, however, it is true that an otitis media, in the acute suppurative attack, is always associated with pain, and is remotely associated with some phase of naso-pharyngeal disease. These exceptional cases are found in phthisical patients, or in certain cases of chronic otitis externa, where there has been ulceration earlier in the history of the case, which in its later stages has passed unnoticed, until the membrana tympani has sloughed, and otitis media has supervened; but this is exceptional.

Symptoms. — The subjective symptoms are those of the Eustachian tube: pain is caused by coughing, sneezing, or any motion of the pharyngeal muscles, as by swallowing or eructation; associated with pain, fever, tinnitus aurium, loss of hearing, vertigo, and, in severe cases, delirium. The objective symptoms are, changes of the membrana tympani,

such as loss of translucency, thickness, loss of the light spot, sometimes a coppery redness, moisture, or a sodden condition. As was said in speaking of diseases of the membrana tympani, myringitis, strictly speaking, is a rare condition; the classification of the disease depending upon the preponderance of the symptoms as regards lesion of the dermoid surface of the membrane, or the internal mucous membrane.

Course. — The course is on to perforation, with relief of the pain; on to suppuration, if not treated. The products of suppuration may escape by the way of the Eustachian tube, in children; and the membrana tympani may not be destroyed. In adults, with unyielding membranes, the latter having been thickened by repeated attacks of catarrhal inflammation, the pus may press in every direction before the drum-head yields, — up to the cerebrum or cerebellum, down to the jugular vein, inward to the labyrinth, posteriorly to the mastoid cells, — and cause serious lesions in either of these directions.

Under treatment, however, the tendency is to resolution, even to perfect restoration. The rapidity, the completeness of repair, is something very remarkable. Some careless people invite suppuration, doing nothing until hearing is gone beyond recovery. This is due in many cases to the popular idea that suppuration is beneficial; such a notion being supported by the relief that follows rupture of the membrane, as well as by the serious results which have followed suppression of the suppurative or purulent discharges, under unwise treatment. The profession itself is not free from blame for giving support to this erroneous view.

Ætiology. — The same causes which produce acute catarrhal disease are active in this form of disease, more especially the use of the nasal douche, sea-bathing, scarlet-fever, measles, diphtheria, and traumatic causes. While it is true that the nasal douche may cause acute catarrhal inflammation, or even chronic catarrhal inflammation, the careless use of this instrument has been followed, in my observation, by acute suppurative inflammation in a large number of cases. The same may be said concerning the practice of snuffing water for the purpose of overcoming obstructions in the naso-pharyngeal tract, as is advised by physicians not aware of the danger to which they thereby expose their patients. The practice of sea-bathing is open to objections for the same reasons, in many cases. In a few instances, inflammation may have been set up by the direct force of the waves in surf-bathing; but, in most cases that have come under my notice, I am satisfied that the salt water reached the tympanum by way of the Eustachian tube, during the violent efforts to free the nose and pharynx from the water. In all these instances, whether produced by the douche, by snuffing, or by sea-bathing, the water acts as an irritant when once within the tympanum, and the inflammatory action passes very rapidly beyond the grade of simple catarrhal inflammation. Scarlet-fever stands first, among the exanthemata, as a cause of middle-ear disease; next, measles; and, third, diphtheria. The changes which are wrought by the first two, leave lifelong traces upon the naso-pharyngeal and tympanic mucous membrane. Diphtheria causes, in my judgment, greater changes in muscular tissue than in the mucous membrane: at least, such has been my view, based upon the cases which I have observed. Phlegmonous inflammation of the tonsils, or chronic enlargement of the same, may act mechanically as a cause; but more has been said and written of chronic tonsillitis as a cause of middle-ear disease than the facts warrant. Traumatic injury of the membrana tympani may act as a cause of acute suppurative inflammation. If promptly and intelligently cared for, such need seldom be the case. Dentition, especially in children, has been shown to bear a direct relation to trophic changes in the tympanum. The same is true in adults; though usually, with the latter, the form of disease will be otalgia neuralgica, rather than otitis. Authorities are now agreed that there exists a relation between the dental branches of the trifacial and the tympanic nerves.

Dr. Woakes, particularly, has given his experience in confirmation of this view; and I take pleasure in referring you to his little work, for a full discussion of this subject.

Diagnosis. — The differentiation between otalgia neuralgica and colic, in children, is, perhaps, the most important point for you to observe. What has been said with regard to suppuration without pain, will lead you to examine patients carefully as regards phthisical tendencies; and the same may be said as to the examination of the teeth, to distinguish between otitis and otalgia. In otitis, the bulging, and evidences of congestion, will be associated with decided loss of hearing; whereas your neuralgic patient will be able to hear normally, unless the case be complicated by middle-ear disease. The angular explorer, used by dentists, should be found in your armamentarium, and will be of great service in the examination of the crowns of the teeth.

Prognosis. — The prognosis of acute suppurative inflammation is favorable under the methods of treatment now in use, but certainly unfavorable if neglected; as its tendency is to the destruction of tissue, and the establishment of a chronic suppurative condition.

Treatment. — The same mechanical treatment which was suggested in acute catarrhal inflammation, is of value in this form of disease. The application of vapor of water, or of hot water to which has been added a few drops of *aconite*, *belladonna*, or *plantago* tincture, dropped into the meatus as hot as can be tolerated, will not only mitigate the pain, but help to abort the disease. If the inflammatory action is not relieved by medication, and if the membrana tympani threatens to rupture spontaneously, paracentesis will not only relieve the suffering, but leave the tissues in a condition more favorable for repair than will be the case if spontaneous rupture be allowed to take place. Any knife similar to a tenotomy knife may be used in an extremity; but either of the forms which have been suggested by Roosa, Knapp, or Agnew is to be preferred. The lance-shaped paracentesis knife[1] will be sufficient to penetrate the membrane; but a

[1] See cuts on p. 22.

knife formed more like a curved bistoury is better adapted, if it is desirable to make a larger incision from the point of penetrating to the periphery. Even a needle, securely fastened in a wooden handle, and guarded within a millimetre of its point by cotton wound firmly upon it, will serve to penetrate the membrane, and relieve the cavity of the tympanum of the accumulated gaseous or fluid contents; and it is astonishing what relief this simple proceeding will give. The fact that a small opening closes very promptly, will necessitate, in most cases, a larger incision; and it is good practice to follow paracentesis by inflation of the middle ear, using Politzer's method or a catheter, as may seem best. Suction applied to the meatus, by the use of Siegel's otoscope, has proved of practical value, drawing the contents of the tympanum through the perforation, or bringing thick mucus through the perforation, so that it can be seized by forceps, and drawn out entirely.

In the transactions of the American Homœopathic Ophthalmological and Otological Society for 1884, W. H. Winslow, M.D., of Pittsburg, Penn., reports a case in which he used Siegel's otoscope in an original manner. Acute inflammation of the middle ear had advanced to such a degree that the brain was slightly involved: incision of the membrane followed by inflation afforded but little relief. Strong suction with Siegel's otoscope caused a flow of bloody serum with immediate relief; and, later, the products of suppuration were removed in a similar manner. The patient made a complete recovery.

Remedies. — *Aconite, belladonna, chamomilla, capsicum, dulcamara, gelsemium,* and *tellurium.* *Aconite* is indicated in high fever, burning skin, great restlessness and thirst. *Belladonna,* less marked redness and heat of surface, less restlessness, but mentally a desire to escape. *Chamomilla* is characterized by the same intolerance of pain on the part of adults, or extreme irritability and peevishness of children. *Capsicum* is of value for adults when the mastoid process is threatened by the inflammatory action. *Dulcamara* and *gelsemium* are of more value in acting against threatening suppuration than when it is fully established. The same symptoms would indicate *hepar sulph. calcarea*, as those mentioned in acute catarrhal inflammation, — the local tenderness about the ear, especially in front and behind the auricle, sensitiveness about the ear, and relief by wrapping. *Mercurius* has the same symptoms as under catarrhal inflammation, but more marked. *Pulsatilla* is especially valuable in the earlier stages of the disease, especially in children. *Tellurium* is indicated in cases where perforation has occurred spontaneously, or where the tendency is to rupture, and to extensive destruction of the tissue; the discharge being of an ichorous, excoriating, and especially fetid nature, smelling like fish-pickle. An intercurrent dose of *sulphur* or *psorinum* will prove valuable in bringing out more clearly the distinctive indications for the previously mentioned remedies. Electricity has proved not only a mitigating agent in the treatment of acute suppurative inflammation of the middle ear, but I am satisfied it has cut short the history of the disease. The same principles underlie its application that have been demonstrated in the treatment of paronychia, — felon. On application of the positive pole to the meatus, covered with sponge or cloth dipped in hot water; and the negative pole to the feet, by means of a hot bath, — and allowing the passage of the current for the space of three to five minutes, great relief is afforded.

This has been true of both the galvanic and the faradic currents, and is commended to those who by experience are qualified to apply this means of relief; as in unskilled hands it is an agent of destruction instead of reparation.

EIGHTH LECTURE.

CHRONIC SUPPURATIVE INFLAMMATION OF THE MIDDLE EAR.

GENTLEMEN, — The remote causes of this disease are the same as, and coincident with, those of acute suppuration; and you will remember, that, in speaking of that disease, I told you that its history, unchecked by treatment, passed on into that of chronic suppurative inflammation. It is also true, that, in a limited number of cases, otitis externa, causing ulceration of the membrana tympani, involves the substantia propria; and, finally, the mucous membrane of the tympanic cavity gives way. Then there is set up a chronic suppurative inflammation, which does not differ in its symptoms, or subsequent history, from one which arises in the tympanum itself.

Symptoms. — The two leading symptoms are the discharges and the deafness. Allow me to call your attention to a term found very frequently in our literature, which is used to designate a disease, — otorrhœa. This is the name, not of a disease, but simply of a symptom of a disease. The make-up of the word, as you understand, indicates its meaning, — "a flow from the ear," — and is no more the name of a disease than is leucorrhœa, which is simply the name of a flow from the vaginal or uterine mucous membrane, or from both. I trust, that in your writings, — as it may be presumed that you will be writers, — you will correct this error. As regards the nature of this discharge, it may be said that it is marked by all the shades of difference between pure pus and a muco-purulent discharge which is more mucous than pus. The purulent discharge may also be laudable, bland, or ichorous, excoriating, sanguineous; and on these small points of difference is based the prescription of various remedies. The deafness is in marked contrast to that in many cases of chronic catarrhal inflammation. And here I wish to mention another error, from which, even to-day, the profession is not altogether free. It is not long since I had occasion to comfort a patient, who had been thrown in a state of great alarm and anxiety by being told by a female physician of some eminence in this city, that the drum-head was destroyed, and hence her hearing was forever gone. This was a serious mistake, as it might have led her to abandon all treatment. The fact which I have noted, of the possibility of greater hearing-power in suppurative than in chronic catarrhal inflammation, will be clear, when you remember the anatomical relations and lesions of chronic catarrhal inflammation. The fact of adhesions and pseudo-anchylosis is a much more serious one, so far as hearing-power is concerned, than articulations bathed in pus, or muco-purulent secretions. So, too, when these secretions are overcome, a great amount of power can be promised by artificial means, as we shall see when studying the adjustment of the artificial membrana tympani or the cotton pellet. There are certain facts touching the calibre and length of the canal, the appearance of the membrana tympani, and of the mucous membrane of the cavity of the tympanum, if the drum-head be perforated, which will, as objective symptoms, help you to distinguish between otitis externa and otitis media. Until the eye is educated, you may not be able to determine whether the granulation, ulceration, pus, or mucus, which is seen, is located in the canal, upon its walls, or upon the walls of the cavity of the tympanum. Indeed, even an educated eye is sometimes under the necessity of employing all the helps which may be brought to bear. Practice, however, will enable you to distinguish the remaining portions of the membrana tympani, if largely destroyed, and will also gradually take in the perspective of greater or less depths of the canal and tympanic cavity.

History. — This is usually one of neglect; and I cannot too strongly deprecate the assurance which is given, not only by friends, but by medical advisers, that *time* will overcome this disease. In fact, the advice which is given by elders to children, and to inexperienced patients, is based upon, and supported by, the advice of the faculty from time immemorial: only within the last few years has there been any thing like an intelligent understanding of the causes of otitis and the reasons why this form of disease leads to fatal results.

Prognosis. — As is intimated by what I have already said, the prognosis is unfavorable if the disease is allowed to take its course. It is certainly favorable under the best instrumental and medical treatment of to-day. I may state here, — not in any spirit of egotism, but to emphasize an important truth, — that, for more than seventeen years, I can review the history of cases, and assert, that where the parents, guardians, or others having the charge of patients, have persisted in maintaining the treatment year after year, in no instance have I failed to reach the desired result. In two cases, children who had suffered from scarlet-fever, barely escaping with their lives, both membranæ tympanorum lost, the suppurative process has been brought to an end; and, by the use of a cotton pellet, these children, now young ladies, are able, with but little difficulty, to understand all ordinary conversation. In one case, however, it required nine years, and in the other, eight, to reach this much-desired issue.

Treatment. — The local treatment is as important as the selection of proper remedies, and I may here repeat what I have previously stated. I do not believe that local treatment in any sense interferes with constitutional, general treatment, but, on the other hand, renders the medicinal treatment more prompt, more successful. I believe, that as in moral matters, so here, "cleanliness is next to godliness." By the use of what is called the "dry treatment," and the application of remedies in triturations, using an insufflator similar to that used in the larynx, the excessive purulent or muco-purulent secretion of the mucous membrane may be gradually modified, until it becomes normal, in no sense suppressed, or giving rise to any thing suggestive of metastasis: on the other hand, you guard against the consequences of destructive processes, which must, in time, reach deeper structures, and bring a fatal result. I have used *sulphate of zinc, bichromate of potash, salicylic acid, borax, boracic acid, alumen ustum, calcarea phosphorica, calendula, nitrate of sanguinaria*, and other remedies, in the first and second triturations, with excellent results; and in no case have I seen unfavorable — immediate or remote — results. In some instances, simple drying of the mucous membrane with pledgets of cotton — removing every trace of the secretion — has been sufficient to stop the secretion; but in long standing, neglected cases, I have resorted to the triturations. You will understand, that in no case is the local treatment to be pursued to the neglect of more important constitutional treatment. Concerning the much-mooted question as to the use of the syringe, I have simply to say, that, for a number of years, I have abandoned its use, save for the removal of foreign bodies, or masses of dry or softened detritus, in the canal or the cavity of the tympanum. When using it, either in clinical or private practice, I follow its use with the absorbent cotton, drying every portion of the exposed mucous membrane as perfectly as possible. I am satisfied that more mischief is likely to follow its abuse than its entire prohibition, in the hands of the laity; and therefore I direct parents, or persons in charge of the case, to dry the ear as perfectly

POWDER-BLOWER.

as possible by the use of absorbent cotton, supplementing their work with my personal attention. In many cases, the presence of granulations or polypi will require instrumental interference. If polypoid growths obstruct the canal, threatening the retention of the secretions, they should be removed by the snare or looped curette, rather than by torsion. If retention of pus is not threatened, the instillation of alcohol, or the application of equal parts of alcohol and saturated solution of boracic acid, will cause the growth to shrivel, and prepare for its easy and painless removal. Its return, however, must be prevented by internal remedies: and the same may be said of excessive granulations; these may be removed by Wolfe's sharp spoon, or by the saturated solution of bichromate of potash, as suggested by Dr. William P. Fowler of Rochester. They usually return, unless combated by indicated internal treatment. But much is secured in the direction of complete symptomology by study of the local indications of a constitutional dyscrasia. As a means of overcoming the unavoidable exposure of the mucous membrane of the cavity of the tympanum, when the membrana tympani has been, to a greater or less extent, destroyed, various devices have been proposed. The first of these was suggested to Dr. Yearsley, an English surgeon, by a patient, who demonstrated to him that he could increase his hearing-power very largely, by the introduction of a roll of paper, which, on touching a certain spot, — which the patient could determine by careful manipulation, — increased the hearing-power so that he could understand conversation — in fact, every sound — with great facility. This led Dr. Yearsley to experiment with balls, or rolls, of cotton; and he demonstrated, that, in a very large number of cases, they acted as a support to the ossicula. This led to the introduction of the artificial membrana tympani by Dr. Toynbee, which consists of a thin disk of rubber, mounted on a silver stem, sufficiently long to reach from the remains of the membrana

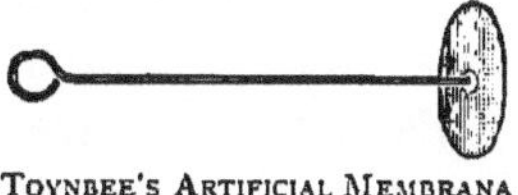

TOYNBEE'S ARTIFICIAL MEMBRANA TYMPANI.

tympani, to the orifice of the meatus. I have found it of most service in those cases in which the remains of the drum-head were covered with a bland secretion, more mucoid than purulent. In very many cases it acts as an irritant, and therefore is not tolerated by the patient. In cases of small perforations, Dr. J. Clarence Blake has suggested the use of disks of sized paper. These, adjusted over the perforations, not only close them, and improve the hearing, but have been the means of promoting cicatricial closure of the perforations. The same is true of the cotton pellet, which in recent years has come prominently before the profession as a substitute for the artificial membrana tympani, and proved to be very valuable in a large number of cases. What is effected by the paper disk is secured by the use of the pellet: it not only protects the perforation, guarding the mucous membrane from the external air, but it serves also to hasten the closure of the perforation. And here I may caution you against the careless removal of crusts which may cover recent perforations; as I am certain that I have retarded the closure of perforations, by too hasty removal of those secretions which nature had thrown out during the formation of cicatricial tissue. The absorbent cotton, rolled loosely between the fingers, and moistened with concentrated petroleum, or even applied without it, is better than cotton saturated with glycerine, as has been suggested by some writers. The petroleum is tolerated; while the glycerine causes a discharge of a lymphy or purulent nature, which is the very thing we desire to overcome. This roll of cotton should be applied so as to support the remains of the manubrium of the malleus, when the perforation is large, or simply to cover the perforation itself if it be small. The following cases illustrate the use of the cotton pellet.

Patients who have themselves made attempts at cleansing the cavity of the tympanum, have at times experienced a loss of sensation and taste on one-half of the tongue, and have been seriously alarmed at the results so produced. You will understand how this has been caused, when you consider the relation of the chorda tympani nerve to the branch of the facial, which passes through the upper portion of the tympanum. In cleansing the cavity, the surgeon may produce this symptom; but it is usually of short duration, and need cause no special alarm. Vertigo may be produced in the same manner, or by the forcible use of the syringe. This is due to direct pressure, either upon the disarticulated stapes or the exposed fenestra rotunda, thereby causing change of the tension of the labyrinthine fluids. This is not usually a serious symptom: but you must not allow this fact to make you unmindful of persistent vertigo, associated with suppurative disease of the middle ear; as we shall see that it is one of the symptoms of necrosis, or caries of the labyrinth.

Remedies. — The indications for the remedies will be found very largely pointed out, by studying the conditions of the naso-pharynx and the Eustachian tube, as well as by considering the objective symptoms which present themselves in the meatus externus. *Calcarea phos.* is of great importance in scrofulous patients with enlarged tonsils and a tendency to grossness of tissue, or to involution of the periosteum, and cancelled tissue of the temporal bone. You have noticed, undoubtedly, that it is frequently prescribed for the poorly nourished children who present themselves at the clinic, those with large heads, large bones, and flabby tissues generally. *Cinchona* has proved of more value in our hands, in cases of hemorrhage from the mucous membrane of the middle ear, than has any remedy laid down in our repertories. This is a matter of clinical experience, rather than pathogenetic knowledge. After having used the various remedies, *cicuta*, *hamamelis*, *elaps*, and *phosphorus*, and failing of satisfactory results in a particular case, I gave *cinchona off.*, in a low potency, on general principles, for the anæmic condition of the patient. To my great satisfaction, the patient's condition not only improved, but the hemorrhage from the ear ceased; and, from the exhibition of this remedy, an improved condition of the tissues began, which was carried to a successful issue by the administration of other remedies. This I believe to be the scope of cinchona, and I use it intercurrent with *calcarea phos.*, *kali muriaticum*, *mercurius*, *psorinum*, *silicea*, *sulphur*, *tellurium*, or *thuya*. *Elaps* is a valuable remedy in the case of children: the naso-pharynx is characterized by dryness of secretion; the mucous membrane of the posterior wall of the pharynx cracks, or is covered with dry crusts; the nares are obstructed, crusty, so that the child has what the old nurse calls "snuffles," and, when sleeping, breathes with the mouth open. The discharge from the ear is thin, somewhat irritating, staining the bedclothes on which it chances to fall, a clear green color. *Hepar sulph. calcarea* is especially indicated in ulcerations, perforations, the particular indications being sensitiveness of the tissues. *Hydrastis canadensis* is indicated by a bland discharge, which is more mucus than pus, associated with dropping, in the posterior nares, of a yellowish catarrhal secretion. *Kali bichrom.* is analogous to *hydrastis*, has the muco-purulent nature of the secretion, but the tissues are more irritable, tending to bleeding or to crusts; and the naso-pharyngeal tract manifests the same disposition. *Kali muriaticum*, introduced by Schüssler, is, in my judgment, a good remedy for excessive granulations; and I have had more satisfaction from its use than from any other single remedy. In repeatedly occurring granulations on the inner third of the canal, about the edges of the perforations, or on the tympanic wall, I always expect improvement under this remedy, in conjunction with the local treatment which I have already laid down. *Kali hydriod.* in saturated solution is of value when there is a certainly recognized syphilitic dyscrasia underlying the local ulceration. *Mercurius sol.* has proved, in clinical experience, to be indicated by this characteristic, — a coppery or metallic odor of the secretion, as well as by the well-known naso-pharyngeal symptoms. *Psorinum* has an extremely fetid discharge, associated with eczematous conditions about the ear, or in other parts of the body. You will compare it with *tellurium* and *thuya*. *Sulphur* is a valuable intercurrent remedy, and may be used when other remedies apparently fail to overcome the conditions for which they are clearly indicated; and its administration often serves to bring out, and render clear, conditions which were before obscure. The general symptoms of the integument should be noted carefully, to guide in the administration of this remedy. *Tellurium* is indicated for conditions of the drum-head similar to phlyctænular conjunctivitis; the whole drum-head appearing dark purple, with elevated spots at various points, which form vesicles, break, oozing a watery discharge, having the odor of fish-pickle, extremely acrid, excoriating the canal, and often the cheek.

The late Professor Carroll Dunham, M.D., made an heroic proving of this remedy, and called my attention to this effect on his own person, some years afterwards. Inspection of the drum-head showed it to have been perforated, and afterwards repaired, the cicatricial tissue being quite extensive. This would argue that the remedy had much deeper action than merely upon the external surface of the drum-head. In fact, this has so proven in long-standing cases, particularly in children.

Thuya is in decided contrast to other remedies mentioned, in that its discharge is bland, thick, the odor being that of putrid meat.

NINTH LECTURE.

CONSEQUENCES OF CHRONIC SUPPURATIVE INFLAMMATION OF THE MIDDLE EAR.

Gentlemen, — You will recollect the statement I made in the introductory lecture, while impressing upon your minds the serious consequences of diseases of the ear. This we shall consider under our present lecture on *Chronic Suppurative Inflammation of the Middle Ear.*

The tendency is to the destruction of the deeper structures. This fact has been noticed by life-insurance companies, so that they have refused to take risks on persons who have been for years subject to a purulent discharge from the ear. The consequences of chronic suppuration are classified by Roosa as follows: —

I. Cicatrices and adhesions.
II. Polypi.
III. Exostoses.
IV. Mastoid Disease.
V. Caries and Necrosis of Temporal.
VI. Cerebral Abscess.
VII. Pyæmia.
VIII. Paralysis.

The dry condition which exists after cessation of a chronic suppuration, is of less serious nature than that which preceded it, even if the hearing be less: better the cicatrices and adhesions, which lessen the hearing, than a suppurative condition, which threatens the life.

Polypus. — This is a term growing out of the old nomenclature, when the nature of the structure was not understood; but it has become so fixed in the literature of the subject, that it is impossible to eliminate it by the substitution of any other term. Studener classifies polypus as follows: —

I. Mucus polypi — Cellular.
II. Fibromata.
III. Myxomata.
IV. Angioma (Buck).

Polypi are composed of connective tissue holding cellular elements and blood-vessels, and are covered by epithelium. Angioma is a purely vascular structure. Malignant growths arising in this region are epithelial carcinoma or osteo-sarcoma starting from the bone.

Treatment. — It is absolutely necessary that there be free exit from the tympanum, for the products of suppurative inflammation: if polypi threaten closure of the meatus, they must be removed by the snare or torsion. The snare is to be preferred; as torsion involves more tissue, and is liable to do mischief. Blake's modification of Wilde's snare is used

BLAKE'S WILDE'S SNARE.

for the purpose. I show you here the modified instrument with paracentesis knife, which can be substituted for the canula. Any portion of the pedicle which remains should be touched with saturated solution of kali bichromicum. Dr. W. P. Fowler of Rochester, a graduate of this college some years since, suggests the use of a saturated solution of bichromate of potash in water; and I know of no safer agent for the purpose. Some months since, having occasion to use bichromate, and not having the solution at hand, I used the red acid solution as prepared for galvanic batteries: this is composed of crystals of bichromate dissolved in sulphuric acid and water, as usually given in the formula with the carbon and zinc elements. Great care must be used in applying this in excess. All the solution should be pressed out from the cotton with which it is applied, and the tissues carefully dried after its application.

Exostoses. — These are bony growths arising from periosteum, and threaten by their growth to close the external meatus: the tissues overlying them should be cut through, and the growth beneath it should be reduced by the use of a burr or chisel. The burr to the dental engine in general use, is to be preferred to the chisel.

Mastoid disease. — It is necessary to observe the distinction between the external periostitis, inflamed glands, and congestion and inflammation of the internal periosteum or mucous lining of the antrum. Dr. Burnett reports a number of cases of mastoid inflammation which were evidently external, primarily; and the condition came to involve the antrum in their later history. These are easily distinguished from those beginning with congestion of the antrum, from the fact that the hearing is not seriously involved, as is the case when the disease is secondary to that of the tympanum. The sensitive lymphatic glands may be recognized, as they are localized, and the surrounding tissue is not sensitive as in external periostitis: internal congestion of the antrum, secondary to otitis media, may occur in acute otitis; in fact, we believe that in nearly all cases of acute disease, the lining of the antrum is more or less involved. A writer in the "Practitioner" reports two classes of acute necrosis of the mastoid complicating otitis media acuta. In these cases, a very brief space of time sufficed to soften the tissues to such a degree that the petrous portion of the temporal bone was laid open with a scalpel, cutting like wet leather. The relief felt by the washing and drainage was immediate; and so perfect was the repair, that six months afterward the bone was as stony as ever. In chronic suppuration of the middle ear, the conditions are such that we are liable at any time to an involution of the mastoid region.

Symptoms. — The local symptoms are redness and swelling of the mastoid process, associated with an increased flow of pus from the meatus, possibly in some cases an entire suppression of the discharge. There is a marked constitutional disturbance, rise of temperature, increase of pulse; and a marked anxiety, or haggard expression, has characterized the cases which came under my notice.

The history is usually a brief one, from the fact that unless relief is afforded a fatal result usually follows.

Treatment. — *Perforation of the mastoid antrum.* The operation is an old one: according to the statement of Roosa, it was made by Jasser in 1776, and by others at various dates. Since then it has fallen into disrepute. Dr. A. B. Crosby made an operation three times with a gimlet, the first being as early as 1864: his three patients recovered.

HEAVY KNIFE FOR MASTOID INCISIONS.

Schwartze's articles served to call renewed attention to the subject, and the operation has been more generally made since that time. Roosa lays down the following rules for the operation: —

"*First*, The integument and periosteum of mastoid process should be freely divided in all cases when there is great pain, tenderness, and swelling in this part.

"*Second*, Such an incision should also be made whenever severe pain, referred to the middle ear, constantly exists, and which is not temporarily relieved by the use of leeches and the warm douche, etc.

"*Third*, The bone should be thoroughly examined by the aid of such an incision whenever we have good ground for suspecting that the bone is diseased, or pus retained in this part.

"*Fourth*, The mastoid process should be perforated after such an incision, whenever the bone is softened; or, if a fistulous opening is discovered, this should be enlarged. It should also be perforated when the suppuration of the middle ear involves the mastoid cells or antrum to such an extent that thorough drainage cannot be secured through the membrana tympani or external auditory canal."

In making the operation, Schwartze uses chisels; and, a free incision having been made down to the bone, he cuts

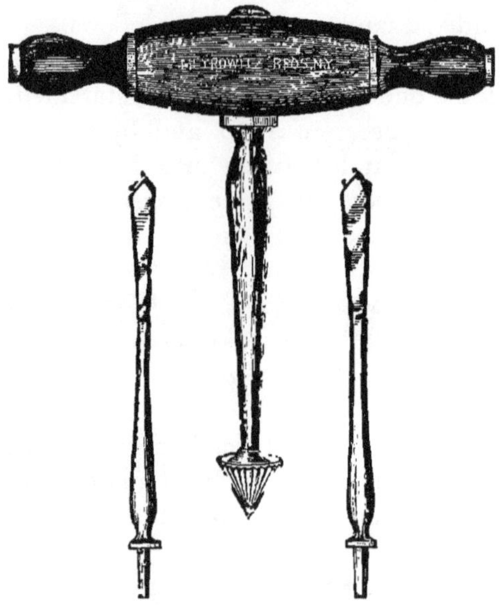

BUCK'S DRILLS.

a small opening to the antrum, enlarging it with gouges until the free opening is secured. Buck's method consists in the use of drills of different sizes to effect the entrance,—one of which is conical-shaped, to enlarge the opening. The point selected for placing the instrument, should be a quarter of an inch back of the external meatus, and on a plane below the level of the upper wall of the canal. A free opening must be maintained by the use of a tent, and the antrum and the middle ear thoroughly washed and drained, that free exit be afforded to all tracts of suppuration. In cases where there has been suppression of the discharge from the auditory canal, previous to the operation, I have noticed that the discharge was re-established sometimes in a few hours after the mastoid antrum was opened. It is undoubtedly true that the engorgement of the deeper portions was relieved by the incision.

Caries of the cranial bones is a more serious matter than in other parts of the osseous system; for Markoe shows that

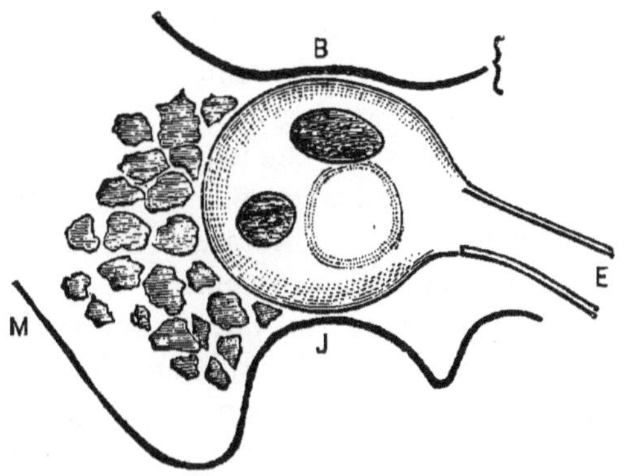

DIAGRAM OF MASTOID CELLS AND TYMPANIC CAVITY.

The inner wall of the cavity is exposed to view, with the round and oval windows and the promontory. M. Mastoid cells. J. Jugular fossa. E. Eustachian tube. B. Base of brain. (A. L. Ranney, from Roosa's Treatise.)

exfoliation of portions of diseased bone is more rare: hence the entire bone is liable to extensive disease. This is true of the temporal, and accounts for the serious results of lesions of the tympanum. The inner wall of the tympanum, its floor and its roof, are very liable to caries: the promontory, being more exposed, may be involved very early in the disease; hence the comparatively frequent exfoliation of the cochlea, or the vestibule and semicircular canals. If the ulceration of the floor of the cavity result in death of the periosteum, caries may expose the Fallopian canal, and cause paralysis of the facial nerve, either from pressure or neuritis: if from pressure, there is a possibility of recovery; as the pressure is resumed by absorption. One case under my care has recovered under internal remedies and the induced current. Caries of the squamous portion of the temporal may prove a serious matter: if its thin tables yield to the advancing inflammation, the periosteum of the internal table being dura mater, a meningitis follows as the necessary consequence.

Cerebral abscess. — Meningitis in acute necrosis may cause extensive effusion, and death be preceded by coma and convulsion. In caries the process is a less rapid one: the molecular death may cause extensive destruction of the dura mater; hence suppuration results, the cerebral tissue breaks down, and abscess is the exciting cause of paralysis of some nerve-function before death of the patient.

Phlebitis is another possible complication of caries of the walls of the tympanum. The literature of mastoid disease contains reports of undoubted extension of the disease to the lesser veins, and even to the lateral sinus. The symptoms are such as are present in inflammation of the veins elsewhere in the body, — induration of the tissues of the neck, involving the direction of the sterno-cleido-mastoideus, and yielding to treatment only as the mastoid disease lessens in force. It may give rise to the more serious condition of pyæmia.

Pyæmia. — By the introduction of pus into the circulation, a systemic poisoning of the blood results: there is a sudden rise of temperature, increase of pulse, chills, and a marked distress manifested in the countenance of the sufferer. Multiple abscesses may result from the septic condition, or localized inflammation of the veins, pneumonia, hepatic abscess, etc., constituting one of the gravest complications possible.

Paralysis. — Failure of the facial nerve has been mentioned as a result of necrosis, or caries of the floor of the tympanum; but lesion of the brain may result in paralysis of one side of the body, and not cause death. Indeed, the possibilities of the explanation of some forms of nervous disease, by the study of co-existing aural disease, is mentioned by recent writers.

Hemorrhage. — Ulceration of the walls of blood-vessels lying, for a portion of their course, in immediate relations with the osseous structure of the temporal, may cause escape of either venous or arterial blood; if the vessels are small, and pressure can be exerted upon them, no concern need be felt: but death has occurred from sudden and uncontrollable hemorrhage.

Remedies. — I am satisfied that remedies do control, not only the primary conditions, which lie remotely at the foundation of the suppurative process, which terminates in caries, but that we may expect repair of the bony structures. This takes place under the continued use of remedies known to modify the repair of bone in other parts of the osseous framework of the body. No possible reason can be advanced why this bone should be the one exception to the recognized action of such remedies. *Belladonna, ferrum phos., gelsemium,* exercise control over the arterial supply, here as well as elsewhere in the economy. *Hepar sulph. calcarea, silicea, calcarea fluorica, Hecla lava,* act upon these tissues to check destruction, as well as to hasten their repair. *Capsicum annuum* would be scarcely thought to have any control over mastoid inflammation, were it not that clinical experience has confirmed the physiological indications laid down in the proving.

TENTH LECTURE.

UNUSUAL DISEASES OF THE MIDDLE EAR.

THERE are some conditions which do not necessarily belong to any division of this classification, of which I will speak at this time. In connection with many cases of what I have termed the post-catarrhal form of chronic catarrhal inflammation, there is a functional condition, which has, as its characteristic feature, sounds which are undoubtedly produced by spasmodic action of the tympanic muscles. These may be snapping, as of the nails of one's fingers, or like the flapping of a bird's wings before the ear. Sometimes these sounds are described as being in the head rather than in the ear. On one occasion, at the annual meeting of the "State Medical Society" of this State, a case of this kind was reported. In some remarks made by the late John F. Gray, M.D., he spoke of a case in which the subject believed these sounds to be produced by spirits, or at least to be connected with manifestations of so-called modern spiritualism. Dr. Gray stated that he was able to convince the patient that this was not the case, by producing similar sounds in his own person at will. This provoked some merriment; and, in remarks made in support of the doctor's claim, I said I saw no reason why the will-power could not be brought to bear to reproduce these sounds, as we know that there are both voluntary and involuntary nerve-supply to the tympanic muscles. I was invited by Dr. Gray to visit him at his own office, on our return to the city; and I did so within a few days, when he demonstrated to my satisfaction his power to produce at will the snapping sounds, which are characteristic of the spasms of the tympanic muscles. On placing my ear immediately to his own, sounds similar to that of snapping one's nails were repeatedly recognized ; and he could produce them on either the left or the right side. Careful inspection of the drum-head did not demonstrate any motion of the same. I regret that I did not use a manometer, to demonstrate any slight motion which could not be recognized by the eye. These functional disturbances of the muscles are similar to those that occur in the muscles of the eye.

Certain organic changes that are unusual should be noted. The presence of blood in the cavity of the tympanum, which is undoubtedly due to the giving way of minute vessels of the mucous membrane, called by Professor Roosa, "otitis media hemorrhagic," is believed to be associated more particularly with Bright's disease. Of tubercular degeneration I have spoken in connection with acute suppurative disease without pain. Desquamative disease of the mucous membrane lining the tympanic cavity of the mastoid cells may occur with an imperforate drum-head, but has been more frequently observed in connection with extensive perforations. It would seem to be in the direction of extensive exfoliation of the epithelium, which accumulation undergoes fatty degeneration, the fatty products changing later to cholesterine. This condition requires careful cleansing, removal by scoop or pipette syringe, so that accumulations and pressure shall not cause the cancellated tissue to break down. Embolism may occur in some of the larger branches of these tympanic vessels : and, in a case recently seen in the Ophthalmic Hospital, I believe this to be the explanation of the condition which exists ; the drum-head, apparently, being changed into a nævus, or mass of venous vessels.

Cancer may invade the temporal bones, as well as any other part of the body, and has nothing of particular note in this locality.

Burnett mentions a case of hairs growing in the cavity of the tympanum and mastoid cells, and a specimen was shown by Toynbee. Dr. Tilbury Fox stated that they were nourished in the cells, and could not have been introduced from without.

The exudation of lymph is the last item to notice, and its nature and effect may be understood by the details of the cases which have come under my observation.

The secretion in the tympanum is a clear, serous fluid, which moves freely in recent cases, so that the line of its level may be seen if the membrana tympani is transparent. When the patient sits or stands, the line is seen horizontally across the membrane; but, on lying supine, the fluid flows into the antrum. Inflation of the cavity of the membrane causes a dissipation of the fluid, and stimulates its absorption. Adhesive inflammation is more likely to follow serous exudation in the middle ear, than simple chronic catarrh: hence it is necessary to watch the after-history, and use inflation to keep the walls separate.

Hemorrhage in the middle ear is mentioned as a complication of Bright's disease. I am satisfied that it may occur independent of renal disease.

Mr. R. S., a young man of feeble general constitution, suffered from chronic naso-pharyngeal catarrh: it extended to both tympana. Each winter he had occasional treatment, but never regained the power lost. The drum-heads became very much retracted; and at last, in one of the acute attacks, the color of the transmitted reflection changed to a dark green, or greenish black. Under treatment, the hearing improved, but the color remained the same. Two years ago this winter (1885), the membrane in one ear bulged, without the usual symptoms of acute disease, such as pain, fever, etc. I proposed paracentesis, but it was not allowed. A few days afterward, the drum-head yielded; and a watery fluid, mingled with dark bloody masses, flowed for a few days. The stuffy sensation, which he had complained of, passed away, and the discharge ceased. Afterward the drum-head presented an irregular appearance, being bulged at various points over the entire field of view. The left membrane was similar in color, but not lobulated.

This fall (1884), the right ear began a similar course, feeling stuffed; and the hearing was diminished more than usual. Examination showed two very marked projections at the superior junction of *Mt.* and meatus; the bulging was very great; and the external layer of the membrane was whitish, overlying the dark-blue beneath. Any operation was dreaded and refused. After a few days, the tissues yielded, and a similar dark bloody discharge appeared. The projections had receded, but the surface of the drum-head was more irregular than before. It is a matter of regret that the subject of disease is so intolerant of examination and treatment. My view of the case is, that there is a dilatation of the venous supply of the tympanum, which is aggravated by attacks of acute catarrhal disease of the naso-pharynx.

Pulsating Tinnitus in Disease of the Middle Ear. — There is one unusual symptom noticed in some cases of middle-ear disease, — a beating sound synchronous with the impulse of the heart. In studying these cases, it will be found, that, in many, the subjective sound will be modified by pressure upon the carotid; and in some instances it has been entirely overcome while the pressure was maintained. While it is true that the existence of middle-ear disease acts to make this form of tinnitus more seriously annoying, I am satisfied that we must be compelled to admit the explanation offered by Dr. Woakes; viz., that the cause is a lack of vasomotor power, and the impulse of the volume of blood is communicated to the vessels of the internal ear quite as forcibly as to the temporal bone as a whole. The cause of the tinnitus is to be found, not so much in the lesion of the middle ear, as in the lack of power in the centres of the sympathetic system of nerves.

Hydrobromic acid exercises a control of this condition of the nervous system, and may bring about a cure, as in the following case: —

Miss S., aged thirty-seven. April, 1883. Has had naso-pharyngeal catarrh for over five years; subject to rheumatism; has been under treatment by a specialist, who relieved the rheumatism, but not the catarrh; H. D., watch 23-240 right, 10-240 left; has to exercise great care in hearing pupils in school; can converse readily with single ones if she can see the face; subject to tinnitus, ringing and beating; the beating is a regular pulsation, in keeping with the action of the heart; right *Mt.* opaque, left *Mt.* fairly clear; Et. dilatable; tissues of pharynx thick and red, very irritable. A mild current of the induced electricity improved the hearing, and reduced the tinnitus temporarily. Various remedies were used with little avail. In June she was put upon the use of *hydrobromic acid* each day on returning from school, and has done so, as the symptom required, till the present time (1885), with complete relief of the pulsating tinnitus. *Kali phosphoricum* has been serviceable in relieving the ringing sounds.

Diseases of the Middle Ear. — Cases.

Case No. 8. — Nathan Mars, aged twenty-three. *Otitis media catarrhalis acuta.* Two weeks ago, by an exposure he took cold. A sense of heat, itching, and burning occurred in the left ear, followed by deafness. Associated with the trouble in the ear is a sore throat, a sense of roughness of the throat, with increase of mucus, and a slight cough. There is also accumulation of mucus in the nares.

This is a typical case of acute catarrhal inflammation of the middle ear, and in direct contrast to a previous case. Inspection of the drum-head shows that it is sunken, together with hyperæmia, especially at the peripheral margin, and about the attachment of the handle of the malleus. The Eustachian tube is closed. It is necessary in this case to get the tube open, inflate the tympanum, thereby dissipate the mucus, and relieve the approximated surfaces. Should these surfaces remain in contact, an adhesion would occur. Remedies must be administered to correct the existing conditions. Such cases as these have been maltreated by aspirating the cavity through the drum-head. By this, you accomplish nothing: the edges approximate, the opening of the Eustachian tube is not facilitated, and the danger of suppuration is increased.

Case No. 13.—Jane O. Shay, aged fifty-two years. Has always enjoyed good health until one year and a half ago, when she was attacked with catarrh of the nose and pharynx. Six weeks ago she was taken with a disagreeable association of noises in the ear: this has been less annoying during the last two weeks. Day before yesterday (Nov. 7) she took cold, causing soreness and deep redness of the throat, an aphthous condition of the mucous membrane of the fauces: the inflammation extended to the ear, causing severe pain. She was placed under the influence of *merc. deut. iodide*, and to-day is much relieved. The condition has been subacute, tending to chronic: the exposure carried it on to the acute stage, and this is the present state. Please notice the appearance of the membrana tympani, and examine the fauces.

Case No. 14.—Robert Gregg, aged eight years. He was attacked on the 25th of last month (December) with deafness, for which he has been treated, and steadily improved until last night. The tympanic mucous membrane was in better condition, and the naso-pharyngeal catarrh had improved; last night the patient complained of pain; and the case is to be classed as acute catarrhal, liable to pass on to acute suppuration. The membrana tympani is bulging by the pressure of the accumulation in the tympanum. The mucous membrane of the tympanum in the upper part of the cavity forms a duplicature upon the body of the malleus and incus in such a way that *cul-de-sacs* lie below the line of the superior wall of the meatus, and secretions may collect in these "pockets." Von Troltsch has specially described them, hence they are often called "pockets of Troltsch." In a similar case occurring in the person of the daughter of one of our physicians, the inflammation caused bulging at the superior border of the membrane, which looked like a large mass of cellular tissue. Firm pressure upon this with a cotton-holder, covered with a tuft of cotton, emptied the pocket, by forcing its contents backward into the cavity of the tympanum, and thus prevented perforation. In this instance the attempt aborted, from the fact, that, being a chronic case, the tissues were unyielding. By pressing the speculum downward so that its inner end will press on the roof of the canal, and looking from below upward, you will the more readily bring the upper portion of the membrane into the field of vision, and thereby notice the abnormal relations of the membrane, of which I have spoken. In some cases, it is necessary to simply puncture the bulging portion, and, with Politzer's method of inflation, evacuate the contents into the canal: you thus relieve the pressure, and prevent a large slough. In others, where this condition has not been recognized, and the slough prevented, a permanent opening exists between the canal and the tympanum. Undoubtedly these may have occurred in some of those cases supposed to be congenital. I agree with the view that there may be a permanent opening between the cavity of the tympanum and the meatus which is congenital (Rivinian foramen). In the person of a medical gentleman of this city, there exists an opening, which is undoubtedly the result of suppurative disease in childhood. Dilatation with a pledget of cotton, the insertion of a cotton tent saturated with *vaseline* mixed with *boracic acid*, and the internal administration of *kali phosphorica*, were effective in checking the suppurative process.

Master J. D. R., aged ten years. *Otitis media catarrhalis acuta.* April 1, 1876. Is a member of a catarrhal family. He had discharge from the ear last winter, and for several weeks has had occasional ear-ache on the left side. The meatus is normal; *Mt.* depressed; tissues of the pharynx hypertrophied. *Baryta mur*, for snapping noises in the ear when swallowing; ear was relieved.

April 15. Took cold, and was confined to the house for a few days. On examination, R. *Mt.* was found bulging and red, L. oozing a lymphy purulent secretion. *Hepar sulph.*

By keeping the tissues perfectly dry, the lining of the inner third of the canal, both R. and L., exfoliated, leaving *Mt.* on both sides congested, irregular, and adherent from previous suppurative inflammations.

1881. In April he had a similar attack. Under the same remedy, followed by *kali mur.*, he made good recovery.

May 4. H. D. 20-20.

May 15. After a slight ear-ache, severe inflammation set in, and the child was confined to the house for ten days. Under *hepar sulph.* and *belladonna* the high inflammatory action was controlled, with perforation of the membrane.

1882. In February a similar attack confined the lad to the house for two weeks. The remedies then used were *ferrum phos.* and *chamomilla;* perforation did not occur.

March 25 to April 13 he was under *kali mur.;* H. D. rose steadily to 20-20.

April 28. Catarrhal influenza set in, which was relieved by *euphrasia*, and for discomfort in the ear was put upon *ferrum*, which relieved the condition.

June 6. He hears 20-20.

April 23, 1883. Has remained well up to the present time; *Mt.* adherent, but hears 20-20.

Mr. William E., aged thirty years. *Otitis media catarrhalis acuta*, with rupture of the *Mt.*, and ecchymosis in same.

Oct. 9, 1873. Had been under treatment during the previous year for suppuration of the L., ulceration of *Mt.* and walls of canal; removed cellular polypus; ulcer healed; H. D. 12-240; now suffering from acute bronchitis; the severe paroxysms cause pain in the R. ear; the right *Mt.* is congested, and H. D. only 6-240; it had been 20-20 when last tested. *Gelsemium* relieved the pain, but the cough was very severe. *Bryonia*, *causticum*, and *rumex* in succession failed to relieve, and Oct. 14 the *Mt.* was ecchymosed at various points. A small clot of blood was removed from one spot. *Cotyledon umb.* finally relieved the cough. The congestion gradually faded as the hearing improved.

Dr. F., aged fifty-five years. *Otitis media catarrhalis acuta.*

Dec. 16, 1871. Dec. 10 had an attack of influenza; pain in left severe, followed by discharge of yellowish lymph; total loss of hearing; pain has gradually ceased; H. D., watch, right, 4-20; left, -20; *Mt.* and inner third of canal diffuse inflammation, dermoid layer exfoliated; when removed, tissues red and moist; after several attempts, Politzer's method inflated the tympanum; H. D. right, 5-20; left, c-20. *Mercurius vivus* was advised, and occasional inflation.

Dec. 22. Better; H. D. after inflation, right, 15-40; *Mt.* clearer. *Merc. v.*

Dec. 26. Relapsed on account of exposure in the practice of his profession; hears watch on auricle; white shreds over *Mt.;* made an incision through *Mt.;* Politzer; perforation whistle; no gain; no pus in tymp. *Mercurius.*

Dec. 28. Better; *Mt.* was healed next morning; now more clear, folds can be distinguished; Pol. +; Siegel's speculum gives good degree of mobility; itching in ear; H. D. 9-240. *Sulphur.*

Jan. 5, 1872. Better; H. D. right, 10-20; left, 10-240; *Mt.* moves more freely. *Sulph.*

Jan. 26. Improved; H. D. right, 10-20; left, 30-240; *Mt.* clear, and good degree of mobility.

May 17. H. D. 8-20 right and left; *Mt.* clear, but slightly depressed.

Mr. R. M. P., aged sixty years. *Otitis media catarrhalis acuta.* Oct. 2, 1880. Since two weeks he had suffered from head-cold, which caused severe pain in right ear, with subjective noises; *Mt.* thick and reddened; Et. closed; H. D. c-20 right, 10-20 left. *Ferrum phos.*

Oct. 4. Ear feels clearer; *Mt.* less red; Et. closed. *Merc. dulc.*

Oct. 8. Removed exfoliated epithelium; H. D. c-20; *Mt.* less red. *Merc.*

Oct. 13. Better; Et. dilatable; *Mt.* clearer after inflation; H. D. 2-20. *Merc.*

Oct. 20. Improving; removed epidermis thrown off; *Mt.* clearer; H. D. 3-20; Pol. +, then H. D. 3-20. *Merc. d*

Nov. 10. Patient thinks he hears as well as formerly; *Mt.* fairly clear, but thick; inflation easy; H. D. 10-20 right and left. Probably the hearing of this patient had been below standard previous to this acute disease.

Case of W. H. H., aged thirty. Grocer, blond, fair degree of health and strength. Has slight naso-pharyngeal catarrh. Sept. 12, 1871. Has had no ear-trouble, to his knowledge; one week ago took cold, and became gradually deaf in left ear; has been under care of homœopathic physician without relief of deafness; H. D. R. 20-20, L. -20; R. *Mt.* normal, L. *Mt.* uniform; red color, glazed appearance, depressed, bones prominent, folds deep; Politzer's method used, then H. D. L. 10-240; *Mt.* less depressed, and the entire surface covered with rings, formed by bubbles of mucus in the cavity of the tympanum; repeated the inflation, then H. D. 14-240, and bubbles change their relations; third inflation, H. D. 22-240; patient has no pain, but a sense of fulness joined with a sound as of bubbles bursting in the ear. *Graphites.*

Sept. 13. Better till this morning, when the ear felt closed; H. D. R. 20-20, L. 6-240; *Mt.* depressed and glazed as before; Politzer's method +, then *Mt.* less depressed, and inner surface covered by rings as at previous inflation; second trial, then 22-240 sound similar to that heard at last visit; he has bad odor from mouth and nose, and is subject to an itching eruption every summer; feet cold and sweaty. *Sulphur.*

Sept. 14. Better; Valsalva's experiment +; H. D. 30-240; Politzer +, then 33-240 before inflation; *Mt.* looked more normal, having lost its glazed red look, and was less depressed; after inflation, there was the same appearance of rings, but the bubbles were fewer and larger.

Sept. 16. Much the same; Val. +; R. normal, L. H. D. 35-240; Politzer's method +, then 50-240; *Mt.* before inflation depressed, but more normal opalescence; after, reddened over upper half and along line of manubrium; no sign of bubbles in cavity of tympanum. No medicine.

subjective noises. The existing trouble, however, he is inclined to date as far back as four years ago. Eighteen months or two years ago he began to notice failure of hearing: this remained stationary for a time. About a year ago the subjective noises of which he complained began: he described these as a hissing. Examination shows that the membrana tympani is depressed, somewhat opaque, due to previous congestions. The naso-pharyngeal mucous membrane is secreting excessively, a whitish, opaque mucus, and is follicular or granular; the Eustachian tube of the right side, closed. This is the brief history of a large number of similar cases. I will treat Mr. R. before you in order to show you how inflation is performed. As I have said to you, Valsalva's method of inflation is an experiment simply, and should not be used as a method of treatment. Yet it is advised by many physicians who are ignorant of its danger and unhappy results.

Some aurists may have been so imprudent as to do likewise; but I think that most know better, and advise against it.

I will use the Eustachian catheter first on the right side, as you see, and succeed in forcing the air into the middle ear. We will now use Politzer's method of inflation, introducing nose-tip into the opposite nostril; as, in catarrhal thickening of the nares, the air passes more readily to the ear, than when it is introduced on the same side. Inflation frequently has the effect of decidedly diminishing the subjective sounds for the time being, if it does not entirely overcome them. If it simply modifies them, then we may follow the inflation by traction or suction with Siegel's pneumatic otoscope. The membrana tympani being drawn outward, the adhesions which exist in the cavity of the tympanum are subjected to tension, and in some cases the sounds are entirely removed. These distressing sounds in the ear are caused by the pressure of the air upon the membrana tympani, without a corresponding pressure of the air in the Eustachian tube, and cavity of the tympanum, air being excluded by a closure of the former: hence the stapes is forced inward; and the tension of the labyrinthine fluid is so charged, that the vibrations of the fluids cause an irritation of the terminal filaments of the auditory nerve. The hypothesis that, to my mind, best explains this subjective condition, is, that the vibrations caused by the flow of the blood through the mechanism of the internal ear are subject to, and modified by, this abnormal interlabyrinthine pressure. These sounds are often our first evidence of what is going on in the middle ear, and much can be done to check the abnormal process which causes them. When anchylosis of the stapes has not occurred, but the tympanic muscles are in an enfeebled, and possibly atrophic, condition, much may be done by the use of electricity. If the tympanic muscles have been in a state of disuse for a great length of time on account of catarrhal thickening, this disuse results, as in all similar cases, in loss of power. If it has existed but a short time, it is readily restored; but, if continued for many years, the bellies of the tensor tympani and the stapedius muscles undergo fatty degeneration, and possibly form adhesions in some portion of their relations to other parts. The action of electricity is undoubtedly the same upon muscular tissue here, as elsewhere in the human body.

The galvanic current will affect tissue profoundly in the way of nutrition, and arouse the dormant energies of the motor nervous supply. Applications of the Faradic or induced current increase muscular activity, and overcome the atrophic condition of the muscles themselves. There is much scepticism on the part of members of the profession in regard to the use of the electric current in diseases of the ear. Many European aurists, however, as Brenner, Field, and Woakes, are strong advocates of its use. It is a matter worthy of remark, that those who have used it the least are most sceptical concerning its value; while those who are acquainted with its sphere of action, by oft-repeated applications, are most enthusiastic in its favor.

Mr. S. C. B., aged fifty-five years. *Otitis media catarrhalis chronica.* March 4, 1874. Had naso-pharyngeal catarrh, according to his own statement, five years, but undoubtedly longer. Canal is normal; Eustachian tube dilatable; *Mt.* thick, but fairly movable; H. D. right, contact-20; left, 1-240. *Kali hydriod.* Improved steadily under this and *iodide of baryta* until May 27, when H. D. in right was 30-240; left, 40-240. He then omitted treatment for a month, and relapsed; was under treatment until July 22, when the hearing had advanced to 3-20 right and left. He omitted treatment again for a month, and relapsed to 18-240 and 21-240; under treatment he again improved.

Dec. 2. H. D. right, 28-240; left, 40-240. Omitted treatment until 1876. Dec. 6. H. D. 2-240 right and left. Under treatment with *mercurius dulcis* and *baryta mur.* till Feb. 7; 7-240, right; 8-240, left. Was not seen again until Oct. 2, 1877, when his hearing had fallen to 2-240 and 4-240. Under treatment at intervals until Feb. 18, when H. D. was 9-240 and 10-240. Omitted treatment until the fall of 1878, when H. D. was 5-240 and 6-240. A few treatments sufficed to bring it up to 6-240 and 8-240, when he became completely discouraged, and abandoned treatment. This case illustrates the necessity of continuous treatment in order to preserve even a fair degree of hearing-power in advancing years. It is to be noted, that, although the improvement was marked in a year, the relapses were more decided, and the degree of restoration much less. I am convinced, by successes secured in similar cases, that continuous treatment, even at more prolonged intervals, will conserve the hearing-power to a very large degree, but that spasmodic treatment is very unsatisfactory.

Mrs. G. G. H., aged thirty-five years. *Otitis media catarrhalis chronica.* April 6, 1881. Since six months has suffered from prostration after lung-fever, then confinement, with very tardy convalescence. Both ears were affected, without pain, with subjective noises; meatus normal; *Mt.* depressed; excessive naso-pharyngeal catarrh; Eustachian tubes dilatable; there are symptoms of specific disease. The child was short-lived. The mother lost nearly all her hair. The history of the case, as given me by the family physician, was somewhat obscure; and he was unable to make it clear, even to his own satisfaction. H. D. 19-240 right and left. *Strychnia* and mild Faradic current.

April 11. Better; head stronger; noises less; H. D. 28-240.

April 18. Better; head still stronger; noises still less; H. D. 33-240.

April 27. The same; H. D. 46-240, right; 38-240, left.

July 19. Has continued to improve until this date under Faradism after May 9. Received *ferrum phos.*, *baryta mur.*, and *sulphur* for varying symptoms of the naso-pharynx; H. D. 20-20 right and left.

I have seen this patient occasionally since the above date, using Faradism and various remedies for the subjective noises with relief, but have never succeeded in completely overcoming the hyperæsthesia of the auditory nerve.

Miss A. F., aged thirty years. Jan. 8, 1877. Since one year has been troubled with catarrhal disease affecting both ears. After an acute attack, has ringing in both ears; appearance nearly normal, except the *Mtt.* depressed; she has naso-pharyngeal catarrh; H. D. right, 6-240; left, 6-20; under *verbascum* for a catarrhal cough worse at night; *mercurius dulc.* for a relapse, increase of noises; *chininum*, *sulph.*, *phosphorus*, *baryta*, *arsenicum*, with the continuous current, up to the 22d of March, and the secondary current up to the 3d of July; H. D. improved to normal; subjective sounds very slight indeed.

Mrs. M. S. G., aged thirty-five years. *Otitis media catarrhalis chronica.* Nov. 16, 1878. For two weeks has had trouble with the right ear from cold; the right side of the neck is infiltrated, and all the region about the meatus anteriorly and posteriorly very sensitive to touch; the canal is normal; *Mt.* depressed; naso-pharyngeal catarrh; Eustachian tube closed; H. D. 3-240. Her family physician had prescribed the internal remedies, and desired her to have the benefit of such mechanical treatment as was indicated in the case. Under inflation by Politzer's method, the use of a mild galvanic current for the first week, followed by the secondary current, she improved rapidly until Dec. 28, when H. D. was 20-20 right; H. D. in the left ear was unusually acute, giving a record of 30-30.

Miss E., aged twenty-four years. *Otitis media catarrhalis chronica.* May 11, 1874. For two days has had trouble with both ears; sudden loss of hearing, without subjective noises; *Mt.* depressed, thin; has follicular pharyngitis; Eustachian tube dilatable; H. D. right, 27-240; left, contact. *Kali hydriod.* Relieved by inflation, Politzer's method.

May 16. Subjective noises have set in; H. D. 10-20 right and left; after inflation, right, 20-20; left, 16-20. *Cotyledon umb.*, to relieve the bubbling sound in the right ear, as though there were mucus in the tube.

May 21. Subjective noises; H. D. 16-20 and 12-20; after inflation, 18-20 and 14-20; itching deep in the ear, as though in the Eustachian tube. *Causticum.*

May 27. Subjective sounds have ceased; H. D. the same as at last date; after inflation, 18-20 and 16-20.

Oct. 12. H. D. 18-20 right and left; has remained free from subjective sounds since previous visit.

Miss M. S. D., aged thirty years. *Otitis media catarrhalis chronica.* May 2, 1877. Has had naso-pharyngeal catarrh for years; gradual loss of hearing, without subjective sounds. Dr. Knapp gave an unfavorable prognosis. *Mt.* depressed; Eustachian tube dilatable; tissues of pharynx hypertrophied; H. D. 4-240 right and left. Mild galvanic current, as before suggested; the negative pole at the trifacial, the positive on the tongue. *Kali hydriod.*

May 16. Improving; H. D. right, 6-20; left, 12-20.

June 1. Continued under the same remedy, with mild applications of the Faradic current, following the galvanic current; had one slight relapse, affecting the ear only, and has heard 20-20 since the 25th ult.

Miss C., aged twenty-five years. *Otitis media catarrhalis chronica.* April 7, 1877. Since two years has had catarrh;

H. D. right, 7-240; left, 18-20; meatus normal; *Mt.* depressed; Eustachian tube closed; pharynx granular. Received *mercurius.*

April 30. Right, 10-20; left, 20-20.

May 28. Better under *kali hydriod.;* improved steadily; H. D. 14-20.

June 4. H. D. 20-20; yawning, sneezing, clearing the throat, and any other muscular action of the pharynx, cause snapping, cracking in the ear. Relieved by *gelsemium.*

Mr. N. M., aged fifty years. *Otitis media catarrhalis chronica.* June 6, 1878. After the removal of an accumulation of cerumen from the canal, the hearing was very dull. *Mtt.* congested; H. D. right, 24-240; left, 44-240. Under the action of *mercurius dulc.*, with inflation by Politzer's method two or three times a week, on the first of July the patient heard 18-20 right and left. I had to see the patient occasionally in later months; have heard from him, through friends, up to the present season, and understand that the benefit has been a permanent one.

Application of the secondary current.—Mrs. C. A. A., aged thirty-one years. Jan. 10, 1880. Eleven years ago had abscesses in both ears. Recently, ringing has set in, in both ears; hears with right ear, 5-240; left, 7-20; meatus normal; membrana tympani adherent; pharynx granulated; Eustachian tube dilatable. Under *kali hydriod.*, and the application of electricity in the same way, the patient improved so that, at the close of treatment, on Feb. 11, 1880, she hears with the right, 10-20; left, 18-20.

Mrs. B., aged forty years. *Otitis media catarrhalis chronica.* April 23, 1877. For seven or eight years has noticed a gradual loss of hearing on both sides; now troubled by low roaring, especially annoying when quiet. The membrane is depressed; the tissues of the naso-pharynx thin, atrophied; Eustachian tube easily dilatable; hearing, on the right side, 1-480; left, 14-240. She has received *mercurius dulcis.* Treatment continued through April and May, up to the 9th of June. Constant current, from 5° to 10°.

June 9. H. D. right, 18-240; left, 6-20. During the month of May, she was under *kali hydriod.*

Master W. O., aged sixteen years. *Otitis media catarrhalis chronica.* Oct. 21, 1873. Seven years since, had scarlet-fever; has been losing hearing gradually during the last six months, with subjective noises; H. D. right, 3-240; left, 1-240; meatus is normal; Eustachian tube dilatable; *Mtt.* depressed; tissues of the throat thick, but not granulated. Under *kali hydriod.*, with occasional doses of *mercurius*, for acute conditions.

Nov. 11. He hears 20-20 right and left.

Jan. 8, 1874. He had omitted treatment for nearly a month. Under treatment till April 15, when the hearing was 20-20 again.

May 18. Acute inflammation set up in the right ear, caused by a carious tooth; H. D. 3-240; the tooth was extracted, and he was under treatment.

July 10. H. D. 14-20; was not seen again until April 11, 1875, when he had relapsed, and had had ear-ache the previous week; H. D. right, 3-240; left, 1-240.

Although the parents of this child were wealthy and intelligent people, they abandoned treatment because the result obtained was not a permanent one; although the relapses were clearly traceable to the causes above noted.

Miss R., aged thirty years. *Otitis media catarrhalis chronica.* April 8, 1878. For two years and a half has suffered from subjective noises in the left ear, caused by successive colds; H. D. right, 16-20; left, 14-240; drum-head not especially abnormal in appearance, except depressed; Eustachian tube dilatable; in the naso-pharynx, granular pharyngitis. *Mercurius dulc.*, with occasional inflations.

April 29. Has improved steadily, except for the voice; H. D. right, 20-20; left, 18-20.

June 1. Subjective noises very much relieved; H. D. for the watch was normal, and for the voice very much improved. This case is one of those exceptional ones, in which the hearing for the voice does not gain to a degree parallel with that for the watch.

Mr. E. W. C., aged sixty years. *Otitis media catarrhalis chronica.* Oct. 30, 1876. For years has had naso-pharyngeal catarrh. Has suffered now for four weeks from an attack of suppurative disease; the usual acute symptoms; H. D. right, 1-240; meatus normal; *Mt.*, which was oozing a muco-purulent secretion at the time he was visited at his house, is now thick and retracted; follicular pharyngitis with a decided ozænic odor. *Mercurius dulcis.*

Nov. 14. Has been under this remedy until this date; H. D. right, 14-240; left, 38-240; his voice sounds strangely to himself, as though passing through the ears. *Causticum.*

Nov. 24. Improving; H. D. 42-240 and 46-240; troubled by a rattling cough. *Sambucus.*

Dec. 2. Improving; H. D. 5-20 and 8-20; cough relieved. *Mercurius dulc.*

Jan. 22. H. D. 10-20 and 14 20; Politzer +; H. D. 18-20 right and left.

Sept. 3, 1881. Has been well until this date; from exposure, had acute catarrhal inflammation of the naso-pharynx, which involved the left ear; walls of the meatus red; *Mt.* congested; roaring in the left ear. *Ferrum phos.*

Sept. 15. Seen twice; is much better; hears well enough to suit himself.

Oct. 6. Not as well; subjective sounds in left ear increased; applied mild Faradic current, the positive pole to the trifacial, the negative to the tongue. *Mercurius dulc.*

Oct. 10. Noises still troublesome; less after application of the galvanic current. *Kali mur.*

Oct. 27. Subjective sounds ceased; meatus normal in appearance; *Mt.* free from congestion; has been under *kali mur.* until this date. Although this patient did not have normal hearing for the watch, he is well satisfied with the results of the treatment.

Mr. R. S. S., aged twenty-five years. *Otitis media catarrhalis chronica.* Nov. 29, 1876. Had scarlet-fever when three years old, when the ears discharged. Since then he has had occasional attacks with subjective sounds, and the hearing has lately grown quite dull from repeated colds. H. D. right, contact-20; left, 8-240; canal very narrow; drum-head not much thickened; tissues of the pharynx hypertrophied; Eustachian tube dilatable. Under *mercurius dulc.* for the acute condition, and *baryta mur.* for the subjective sounds caused by swallowing, yawning, or any other muscular action of the pharynx, there was improvement; H. D. rose to right, 10-240; left, 12-240. During four years the young man was away at college, and visited me occasionally. During the holiday vacation of 1879-80, the irritability of the throat was increased to such a degree by an acute cold, that even slight pressure upon the larynx caused spasmodic action, so that it seemed as if the patient would strangle. This was relieved by *lachesis.* During 1880 and 1881, the acute conditions of the mucous membrane were relieved by *mercurius dulc.*, the spasmodic conditions by *lachesis* and *baryta mur.* In June, 1881, I received a letter from the patient, describing a spasm of the larynx in connection with a cough, which led me to prescribe *magnesia phos.;* and, in reply, he stated that it acted like magic. Since that time he has insisted on keeping that remedy by him. As regards H. D., there is no marked improvement to report.

Electricity. — Dr. P., aged forty-five years. Oct. 19, 1877. In March, 1876, had acute *otitis media,* affecting the left side. Seeks relief from a distressing hissing; hears right, 4-20; left, 10-240. Constant current, positive pole to the left ear, negative pole to the hand on the opposite side, 15°, allowing the current to run five minutes; hears 20-240. *Salicylate of quinine.*

Oct. 22. Right hears 6-20; left, 16-240; no hissing; 15°, constant current; short sitting; right, 7-20; left, 2-20; head clearer. *Salicylate of quinine.*

Oct. 26. Right, 7-20; left, 2-20; after exposure, and loss of sleep, ringing set in again; 15°, 17,000; hissing ceases; right ear, 8-20; left, 20-240.

Nov. 5. Right, 10-20; left, 30-240; hissing sound; 10°, 16,000; hissing ceases at once; hears with right, 16-20; left, the same as before sitting.

Nov. 9. Had cold in the head, which did not affect unfavorably; right, 16-20; left, 20-240; 10°, 15,000; current applied with the positive pole on the tongue, and the negative pole on the trifacial in front of the ear; at the close of the sitting, right, 20-20; left, 37-240.

Nov. 15. Better; right, 20-20; left, 37-240; application in the same manner and the same force; right, 20-20; left, 46-240.

Not seen until Dec. 13. Relapse from cold; right, 14-20; left, 35-240; application of the same force and in the same manner; right, 20-20; left, 52-240.

The gentleman purchased a battery, and uses it himself.

Miss Mary T., aged fourteen years. *Otitis media catarrhalis chronica.* Nov. 24, 1878. Five or six years ago had scarlet-fever, affecting both ears; no pain at present, but subjective noises; H. D. right, 4-20; left, 5-20; the canal is normal; *Mtt.* depressed; tissues of the pharynx thickened; Eustachian tube dilatable; after Politzer, H. D. right, 8-20; left, 30-240. The patient improved steadily, under *mercurius dulc.*, until Dec. 3, then *mercurius protoiodide* until Jan. 4, when H. D. was 20-20.

Mr. C. W., aged twenty years. *Otitis media catarrhalis chronica.* Oct. 2, 1873. From some unknown cause, he has had gradual loss of hearing after ear-ache in the left ear eleven years ago; both ears affected with subjective noises; H. D. right, 3-240; left, 1-240; the canal is normal; *Mtt.* depressed and adherent; Eustachian tube dilatable; tissues of the pharynx thickened. This patient was treated by the passage of a galvanic current through the Eustachian tube and middle ear, by the application of tongue-spatula electoid to the positive pole, the negative pole being applied to the meatus, one per cent solution of the *iodide of potassium* having been injected in the middle ear by Gruber's method; after the passage of the current, H. D. right was 6-240; left, 2-240. *Kali hydriod.* internally.

Oct. 9. He had improved materially; H. D. right, 47-240; left, 5-20; after the passage of the current, and Politzer, right, 6-20; left, 8-20.

This treatment was continued until Nov. 12, when H. D. was 12-20 right and left.

Jan. 5, 1874. H. D. 14-20 right and left; after treatment, 16-20 right and left. This is after an interval of about six weeks. Eustachian tube is dilatable, *Mt.* much clearer; and the result was so satisfactory, that, contrary to my solicitation, the patient refused further treatment.

Oct. 25, 1877. The patient called on me, and stated, that, for the last year, he had gradually failed in power to hear; but he could not be induced to resume treatment. H. D. right, 3-240; left, 6-240.

Mr. A. S. L., aged forty-five years. *Otitis media suppurativa acuta.* Sept. 17, 1873. Has been suffering one week; left ear affected; while snuffing cold water for the relief of catarrh, sudden pain set in, with subjective noises; H. D. R. 18-240, L. -20; right meatus normal; right *Mt.* dull; nasopharynx, catarrhal mucous membrane; left meatus congested; left *Mt.* congested and thick; Eustachian tube dilatable. He was put on *mercurius* with *belladonna* to be taken in case of severe pain. After treatment with Politzer on the 30th of September, he hears contact-20 with left ear.

Oct. 2. He took cold, and relapsed.

Oct. 29. Has improved steadily under *mercurius*, and hears right, 34-240; left, 5-240.

Nov. 12. Still improving; H. D. R. 4-20, L. 3-20. He then omitted treatment.

In 1879 I treated him for facial neuralgia involving the ear, and again in 1877 for facial neuralgia; both times relieved him by *plantago maj.;* the hearing remained about the same, 3-20 right and left.

This case illustrates the mischief which may come from snuffing water, as well as from the nasal douche.

Miss L. E. P., aged thirty years. *Otitis media suppurativa acuta.* Nov. 16, 1875. Has had catarrhal trouble for years, "with occasional ulcers in the ear;" hearing has gradually failed within the last few months; both ears are involved; no pain; subject to noises; H. D. right, 4-20; left, 6-20;

pus in both canals; right and left *Mt.* ulcerated; granular pharyngitis. *Mercurius dulc.*, under which she improved.

Jan. 8. Had intermittent fever; and the discharge of pus increased, so much so, that a so-called "regular" practitioner insisted upon local treatment for the suppurative condition, and touched the suppurating surfaces, both right and left, with nitrate of silver, the result of which was a perforation in the Mt. on both sides; and the patient has come to me in great alarm; H. D. right, 4-240; left, 30-240. By local treatment with absorbent cotton, and occasional inflation by Politzer's apparatus, internal administration of *mercurius*, on the 13th the perforation on the left side nearly closed; H. D. right, 3-20; left, 4-20. The treatment was continued through the month of January.

Feb. 3. H. D. right, 6-20; left, 10-20; perforation in the right ear clear, with a very slight secretion of pus. With some slight relapses, the case progressed until the 24th of March, when both perforations closed; R. *Mt.* was adherent; left free; H. D. right, 4-240; left, 14-20.

Nov. 21, 1876. Opacities clearly defined in both *Mtt.;* H. D. right, 4-240; left, 16-20.

Master F. W., aged fourteen years. *Otitis media suppurativa acuta.* March 3, 1879. During an attack of scarlet-fever, suppurative inflammation set in, involving both ears; *Mt.* bulging, and oozing a thin muco-purulent secretion. The patient was attended at his home daily until the 17th; both canals being dried carefully, and all secretions removed from *Mt.;* both drum-heads perforated, but these perforations closed in a few days.

The patient was treated occasionally until May 23, when H. D. was 20-20 right and left; *Mt.* showing only slight opacities as the result of the inflammation.

Mr. C. S. P., aged thirty-five years. *Otitis media suppurativa acuta.* Feb. 10, 1881. Had trouble with his ears when a child, from scarlet-fever. Suppuration set in a week ago from exposure to cold; the canal, pus in the inner third; *Mt.* granulated; tissues extremely sensitive to the touch.

hepar sulph. Patient was under treatment at his residence, where he was visited during the remainder of the month, making rapid progress under *hepar sulph.*, *kali phos.*, and *silicea*. During the month of March he made rapid progress under *mercurius* and *silicea*, when H. D. right was 15-20; left, 12-20. During the month of April suppuration entirely ceased; the tissues became dry, and somewhat rigid; H. D. right, 3-20; left, 4-20. Under *kali chloridum*, H. D. improved steadily, better for voice than watch; June 20, for the watch, 6-20.

Case 15. — Joseph McE., aged twenty-five years. *Otitis media suppurativa acuta.* Sept. 25, 1882. This patient was taken with suppurative inflammation of the middle ear, complicated by a pre-existing catarrhal tendency. His condition was one of great constitutional disturbance, — severe pain, high fever, temperature 101° and 102°, with corresponding rise of pulse. The auricle and meatus were excessively tender to touch, and the mastoid process red, swollen, and bulging at one point. *Hepar sulph.* was administered, with no relief. The high fever continued, accompanied by severe throbbing of the arteries, especially of the head, for which *ferrum phosphoricum* was given every half-hour. The following day the pulse was reduced in frequency, the temperature was only slightly above the normal, and, upon the mastoid process, the suppuration had raised the tissues above the surrounding surface, to the extent of about fifteen millimetres. The tenderness of the mastoid process had very much diminished, but at this point was so sensitive that it was impossible to touch it with the scalpel. Taking a curved bistoury, and gauging the motion of my hand to take the limits of this projecting portion, I suddenly passed it from below upward through the elevation, opening it completely, and giving exit to a large amount of dark and very fetid pus. This aperture was kept open by the introduction of a tent, and in a few days all tenderness disappeared. The opening was maintained until the tent was forced out by the tissues closing from below; and, before this was entirely closed, the membrana tympani had returned fairly to its function. You will readily notice the tract of the incision, and by percussion and pressure you will detect no disturbance of the integrity of the bone.

This patient recovered such a degree of functional activity, that he was unconscious of any difference of hearing on the two sides, and could see no further occasion for continuing the treatment, which he was urged to do, in order to bring the hearing as near the normal standard as possible.

I am satisfied that the remote results which follow many of these cases can be avoided if the patients can be induced to continue the treatment for a greater length of time after such acute disease.

Case 9.—Patrick B., aged sixteen years. *Otitis media suppurativa acuta.* When crossing the ocean, he received an injury in the ear by being struck with a tin pan. Acute inflammation set in; an abscess, or "bealing" as the Irish call it, followed; and a suppurative process was quickly established. When first seen in the clinic of the Ophthalmic Hospital, the pain, which had not intermitted since the injury, was excessive; pulse and temperature high, the whole side of the head corresponding to the right ear indurated, sore, excessively sensitive, and accompanied by discharge from the ear. The blow with the tin pan acted as a blow upon the side of a child's head frequently does, when given by the parent or school-teacher in a passion of anger. In these cases, the hand is often used cup-shaped, causing a more severe blow as it gathers the air, forcing it into the ear in addition to the direct force of the hand. When a child, cuffed in this manner, is suffering with a pre-existing catarrhal affection of the ear, it is an easy matter to cause a rupture of the drum-head with serious results. There was no doubt of a catarrhal disease existing in this patient's ear before he received the injury on ship-board; and, when struck by the pan, the tissues passed on to a suppurative inflammation, much more readily than they would have done if the middle ear had not been the subject of subacute catarrhal disease. The second day after the blow, an abscess had formed (as the people say). The case has progressed favorably under treatment, so that the discharge has changed from a purulent character to that of a more decided mucous nature, and has lessened very much in amount. The tissues of the meatus are becoming more normal in appearance. The prognosis is favorable. There may result some slight impairment of the hearing. This can be avoided only by care over a somewhat prolonged period, after these acute symptoms have subsided. The treatment in this case has been instillation in the ear of ten drops of *calendula tincture*, and half an ounce each of *glycerine* and *water*. Internally, *ferrum phosphoricum* has been given two days after his admission to the clinic. On the third day the acute symptoms have abated, when *hepar sulph. calc.* was given. Under this the tenderness is less, and the suppuration measurably controlled.

This case progressed favorably, and was dismissed with no perceptible loss of function.

Case 7. — William C., aged thirty-five years. *Otitis media suppurativa acuta.* The right ear has troubled him for three months. The trouble began after an exposure to cold, as a severe burning pain felt more during the night, although not much mitigated through the day. It thus continued for two days, when a discharge appeared, which still continues.

The appearance of the discharge brought relief from the excessive pain; but there have been occasional exacerbations of suffering, although the discharge has continued. Upon an increase of the flow, the patient has suffered less pain. The case has progressed to such a point, that, unless the inflammatory action ceases, the accumulated pus will cause spontaneous perforation of the drum-head, and more or less sloughing will result. As not all the members of the class will be able to examine the patient, I will describe what is found on inspection: A case of acute suppuration of the middle ear, tending to the breaking down of the drum-head. The canal is occluded more or less with pus. Upon drying the canal, as you look inward you see the whole membrane, under the thin coating of purulent secretion, is covered with elevated granulations. These are not healthy, but tend to a chronic condition, and are often the basis of granulation tumors,—so-called polypi. In these cases, internal remedies should be given to lessen the inflammation of the Eustachian tube and middle ear. Inflation of the cavity of the tympanum will separate the membrana tympani from the labyrinthine wall, and bring all the parts involved to a more normal position, thereby modifying the circulation, and in many cases relieving the pain, so that the process of repair is more likely to proceed. Paracentesis as a means of cure is not indicated; because the case has, in my judgment, gone beyond the point in which it would be of value. If the drum-head were bulging at one point, I should consider it advisable to puncture the membrane, and expel the contents of the cavity of the tympanum into the meatus auditorius externus, if it were possible to do so by inflation, or by suction, drawing the contents out into the meatus. But if the membrane is involved through its whole extent, and is transuding the contents of the tympanum, I have found paracentesis to result in an extensive slough; whereas, by keeping the canal and drum-head as dry as possible by the use of absorbent cotton, I usually succeed in preventing loss of the membrane. The remedy best indicated in this case is *kali muriaticum;* and in cases of excess of febrile symptoms, *ferrum phosphoricum*, to be given dissolved in water every half-hour.

Master H. T., aged twelve years. *Otitis media suppurativa acuta.* May 6, 1876. Three weeks ago had scarlet-fever and diphtheria; was taken suddenly with ear-ache, followed by discharge from both ears; right meatus filled with pus; on drying the canal, right *Mt.* is found perforated and sodden; left meatus in the same condition; left *Mt.* in the same condition; Eustachian tube dilatable; tissues of the throat hypertrophied. *Calcarea iod.*

May 10. Much improved; right canal dry; in right *Mt.* perforations closed; tissues thick; in left meatus some pus;

left *Mt.* healed, but injected; condition of the pharynx improved; H. D. right, 41-240; left, 18-240; Politzer, then right, 3-20; left, 4-20. *Mercurius viv.*

May 19. Condition much improved; hears much better; H. D. right, 14-240; left, 10-240; Politzer, then right and left, 8-20.

June 7. Discharge much less; hearing improved; in left meatus, granulations in inner third; right *Mt.* adherent; left *Mt.* thick, yellow.

Continued to improve until July 1, when there were scales of cerumen over L. *Mt.;* H. D. right, 12-20; left, 16-20; Politzer, then right, 18-20; left, 20-20.

June 22. H. D. normal; meatus right and left clear; *Mt.* fairly normal.

Dec. 13. Has suffered relapse in the left ear; left meatus is filled with pus; on removal, *Mt.* is found ulcerated, but not perforated; H. D. right, 20-20; left, contact-20; Politzer, 2-240; excessive catarrhal discharge from nose and nasopharynx, bland, yellow, more mucus than pus. *Pulsatilla.*

Dec. 16. Much better; H. D. 3-240; Politzer, then 12-240; discharge from meatus more mucus than pus. *Pulsatilla.*

Dec. 20. Improving; H. D. 13-240; Politzer, then 24-240; the appearance is the same.

Dec. 23. Improving; H. D. 18-240; Politzer, then 3-20; small amount of pus on the walls of the canal; shreds on *Mt.;* complained of lameness; severe pain in right leg; pain better by motion. *Rhus.*

Dec. 27. H. D. 4-20; Politzer, then 8-20; dry scales on *Mt. Mercurius dulc.*

Jan. 6. Improving; H. D. 8-20; Politzer, then 14-20; canal dry; *Mt.* dry. *Mercurius dulc.*

Jan. 13. Hearing very much improved; H. D. 18-20; Politzer, then 20-20; tissues of the canal scaly; *Mt.* somewhat scaly; lameness somewhat improved, but the cause of more complaint than the ear-symptoms. *Rhus.*

May 14. Left ear is again affected; had severe ear-ache last evening and night; the meatus contains a small amount of pus; *Mt.* is infiltrated, and naso-pharynx engorged and red. *Mercurius viv.*

May 17. Much better; no pain since last date; tissues of the canal dry; *Mt.* dry and depressed. *Mercurius dulc.*

May 24. Both ears very dull; H. D. right and left, 6-20; appearance of tissues much the same; Politzer, then H. D. right and left, 20-20. *Mercurius dulc.*

May 31. Hears as well as ever; H. D. 20-20. No medicine.

March 11, 1879. Has been well until present time. Had slight ear-ache on the night of the 7th inst., and the left ear is discharging; H. D. right, 20-20; left, 6-240; pus in canal; *Mt.* infiltrated; tissues of the naso-pharynx much as on the previous visits. *Mercurius viv.*

March 13. Much better; discharge less; H. D. 17-240; Politzer, then 46-240; pus in canal; appearance the same; remedies the same.

March 15. More pain; soreness of canal; furuncle in the left meatus; *Mt.* not seen. *Picric acid.*

March 22. Nearly well; scales in the meatus, which being removed, showed slight accumulation of pus on *Mt.* *Mercurius viv.*

March 25. As well as usual; H. D. 42-240; Politzer, then 6-20; tissues of canal and drum-head dry. Continue *mercurius viv.*

April 12. Has continued to improve, and to-day reports himself well; H. D. right, 20-20; left, 16-20; Politzer, then right and left, 20-20. *Mercurius protoiod.*, which is all he has had since the last date.

May 5. Found the left meatus wet, without previous pain in the ear; H. D. 16-20; left meatus moist; *Mt.* sodden, and re-covered with exudation. *Mercurius protoiod.*

May 10. The tissues have been kept dry, and the condition has improved very much; soreness below the ear felt on pressure back of ramus of the jaw; tissues of the meatus dry, and *Mt.* dry. *Mercurius viv.*

May 12. Pain last night; discharge again this morning; appearance the same as on the fifth. *Belladonna* and *mercurius.*

This attack continued until June 4, and was relieved by *belladonna* and *mercurius*, and terminated by the formation of a furuncle in the left meatus, which cleared up under *hepar sulph. calc.*

This case illustrates the tendency of middle-ear disease to become chronic after scarlet-fever, and also the fact that furuncle is a frequent complication of middle-ear disease.

Cicatrices. — Mrs. B. M. A., aged thirty years. Jan. 19, 1880. Sixteen years ago had an abscess in the right ear, and recently has had severe pain in the affected ear; H. D. 20-20; meatus normal; membrana tympani is scarred, and, in the posterior inferior quadrant, the outlines of an old perforation are easily distinguished. This perforation does not present the appearance usual when the mucous membrane of the tympanic cavity is exposed. On exhausting the air of the canal with Siegel's otoscope in position, a thin membranous structure is drawn outward into the canal, forming a sac-like protrusion, like bullæ. On forcing air into the canal, this same membranous growth is forced through the membrane into the cavity of the tympanum. There is no evidence of any acute inflammation. *Aconite* given for the neuralgic symptoms.

April 16. The same condition; sensitive about the ear externally. *Hepar sulph. calc.* relieved this.

Master Jacob P. K., aged fifteen years. *Otitis media suppurativa chronica.* April 26, 1881. Over twelve years ago he had scarlet-fever and suppuration; both ears inflamed; no pain, but subjective noises; H. D. R. 10-240, L. contact 240; auricle normal; pus in meatus; Eustachian tube dilatable; R. *Mt.* granular, and L. perforated; pharynx thick. Put on *calcarea sulph.* He continued to improve till Oct. 7, when the granulations had disappeared from the right meatus; and, to my surprise, remnants of *Mt.* were gradually defined, as it separated itself from the labyrinthian wall of the tympanum. A cotton pellet was applied to the perforation on the left side.

From Sept. 2 to Dec. 27, the patient was on *kali muriaticum*, with occasional applications of *boracic acid trituration.* My record shows, that, at the time, R. *Mt.* was fair, but adherent, the left perforation closed by a cotton pellet; H. D. R. 20-20, L. 6-20.

This patient was sent me by Dr. Samuel Lilenthal, who had induced him to undertake the treatment, notwithstanding an unfavorable prognosis from many persons with whom he had advised.

Language would fail to give a clear picture of the mental and physical contrast between the two dates, April 26, 1881, and Dec. 27, 1881.

Miss S. L. M., aged eight years. *Otitis media suppurativa chronica.* Jan. 3, 1880. Five years ago had scarlet-fever, and, after the scarlet-fever, mastoid disease. Has had diphtheria recently, which has aggravated the suppurative disease of the ear, and involved the nares on both sides, extending forward so that the mucous membrane was ulcerated at the anterior openings; H. D. R. 20-20, L. 1-240; the left meatus was filled with pus; the left *Mt.* had a large perforation; pharynx obstructed by hypertrophied tonsils; the odor of pus was such that, at first, I gave *psorinum,* and afterward followed it by *silicea.* I made local applications of *calendula glycerole,* on absorbent cotton, through the nares into the pharynx: this was continued for about six weeks, when the mucous membrane was healed, and the patient could breathe with perfect ease through the nose. She was under observation until June 11, 1881, when the same secretion of the tympanic mucous membrane was more mucoid than purulent, and was free from bad odor; H. D. 6-20. During the latter part of the treatment, the remedies were *calcarea sulph., kali phos.,* and *silicea,* with an occasional dose of *psorinum* as an intercurrent remedy. The results reached were extremely satisfactory to both parents and surgeon. One interesting feature of this case was, the complete mental transformation of the little patient, within two months from the time the treatment was instituted. At the outset, the parents had been compelled to remove her from school: she was stupid, morose, devoid of interest in things which ordinarily occupy a child of her age. She became one of the most active scholars of her class, and was the subject of early promotion in her studies. Language does not convey any idea of the picture which such a patient presents.

Master A. O., aged twelve years. *Otitis media suppurativa chronica.* Oct. 2, 1873. Had scarlet-fever in childhood; after the scarlet-fever, had the first discharge of blood and pus from both canals; ulceration with granulations at the inner third of the canal; right membrane perforated; membrane of the naso-pharynx catarrhal; H. D. R. 7-240, L. 9-240. Under *capsicum*, to relieve the pain in the region of mastoid, and *calcarea iod.*, for the condition of the tissues; hearing improved; the tissues about the *Mt.* became more and more clearly defined, and H. D. rose to 6-20 right and left; but the parents could not be convinced of the necessity of continuing the treatment until the suppurative process could be entirely overcome.

A. H., aged eight months. *Otitis media suppurativa chronica.* April 6, 1877. Since two months of age, she had running from the ears; is a very feeble child; pus in both canals; watery. *Psorinum*, a dose every night.

April 13. Very much better; right meatus clear; left, moist.

April 24. Nearly well; more mucus than pus in left canal.

May 11. Has been vaccinated, and, since that, the secretion is very much worse. *Psorinum.*

May 22. Better again. In the right meatus, cerumen has secreted; from the left, discharge of more mucus than pus.

June 23. Report by letter states that the right ear discharges much less, that the left ear discharges none.

Feb. 7, 1882. Right canal normal; in the left canal, pus and shreds, which being removed, showed perforation in *Mt.* Trituration of *boracic acid* applied locally, *psorinum* internally.

Feb. 23. A perforation clearly defined and moist; more mucus than pus. Blepharitis ciliaris chronica.

At a later interview, the mother stated that the child remained well, with the exception, of course, that the perforation did not heal. I may add that the father died quite recently of Bright's disease.

Mr. W. H. H. B. *Otitis media suppurativa chronica.* Jan. 10, 1873. Has suffered from abscesses in his ears all his life; complains of noises in his ears, with loss of hearing; right meatus, pus in the inner half; left meatus dry; large perforations in both drum-heads; granular pharyngitis; H. D. right, 3-240; left, 1-240; after inflation, 6-240 and 2-240. Under *baryta mur.* and *agaricus* for spasmodic action of the pharyngeal muscles, and *psorinum* and *silicea* for the suppuration. He improved, so that H. D. was 25-240 right, and 20-240 left, and for the voice greatly improved; suppuration ceased.

Master A. E. B., aged three and one-half years. *Otitis media suppurativa chronica.* Nov. 8, 1880. Four weeks ago, had scarlet-fever; both ears suppurating, both canals filled with pus of a dark, fetid character. *Kali phos.* Under this remedy until Dec. 2, when both canals were free, both drum-heads healed; H. D., as far as I can judge in a child of his age, was normal.

April 18, 1881. Has had no trouble since last date; meatus clear; both drum-heads clear; but *elaps cor.* was prescribed for naso-pharyngeal symptoms.

Mr. E. G. B. *Otitis media suppurativa chronica.* Oct. 29, 1877. Had scarlet-fever when a child, and the left ear has troubled him ever since; now the right fails. He had malaria in adult life, and took massive doses of quinine. H. D. right, 1-20; left, 3-240; meatus normal; right *Mt.* adherent; left, ulcerative; tissues defined; naso-pharyngeal catarrh; granular pharyngitis. Under treatment until Jan 22, 1878. Under the action of *mercurius dulcis* and *kali bichrom.*, H. D. rose in right to 10-20; left, 1-20; Eustachian tube dilatable; left drum-head free from pus, but lustreless and thick; naso-pharyngeal condition much improved.

December, 1878. Same condition threatened. H. D. fell to

right, 30-240; left, 2-240. Under the same remedies, the hearing improved.

April 7. H. D. right, 10-20; left, 12-240; Eustachian tube clear. This case illustrates the necessity of attention to these patients during the winter, when, from atmospheric conditions, there is liability of relapse.

Mr. E. A. *Otitis media suppurativa chronica.* Feb. 2, 1875. Has had suppurative disease all his life. In left drum-head extensive perforation; H. D. 6-240. Under *hepar sulph.*, *alumen ustum*, *kali bichrom.*, change in the character of the pus, first, a larger proportion of mucus; and, under *kali bichrom.*, a cessation, even of the mucous discharge, was secured within a year.

Dec. 24, 1881. This gentleman brought his little grandson to me for treatment; and, at my solicitation, he allowed me to apply a cotton pellet to the left ear, with the hope of increasing the hearing. The next day suppuration set in, caused, the gentleman believed, by the pellet. Under *ferrum* and *calcarea sulph.*, suppuration was speedily checked.

Dec. 27. I gave *kali mur.* for the naso-pharyngeal catarrh.

Jan. 10, 1882. A perforation, which occupied a large portion of the posterior inferior quadrant, extending on to the anterior inferior, was entirely free of congestion, but extremely sensitive to touch. The patient declined any further manipulation.

Mr. I. W. C., aged forty years. *Otitis media suppurativa chronica.* (Consequences.) Feb. 3, 1879. Has had trouble for many years, the causes of which were so early in life that he is not able to explain them. He has had advice previously, but no hope of improvement in his condition. H. D. right, 2-240; left, 34-240; meatus normal; right *Mt.* perforated anteriorly; left *Mt.* perforated posteriorly; walls of pharynx thick, red, crypts filled with cheesy accumulations; Eustachian tube dilatable. *Protoiodide of mercury.* Was on this remedy until April 16, when H. D., before inflation, was 12-20 right and left; after inflation by Politzer's method, 16-20 right and left.

April 30. He has been on *baryta mur.* since previous date,

this remedy being given for movements felt in the tympanum during the act of deglutition; H. D. before inflation, 14-20; afterwards, 18-20.

Jan. 14, 1880. He has suffered an accident, having had his foot crushed, which prostrated him; and he relapsed in every condition, so that H. D. right was 6-20; left, 4-20.

Jan. 28. Under *protoiodide of mercury*, H. D. rose at once to 10-20, which he said was, for all practical purposes, sufficient; as the hearing for the voice was much superior to the test as given by the watch.

Mr. George C. S., aged forty-five years. *Otitis media suppurativa chronica.* Nov. 3, 1880. There is a history of suppuration from childhood. For about three weeks has had a slight discharge from the right ear; H. D. right, 18-20; left, 8-240; the left *Mt.* is perforated in the posterior inferior quadrant, the edges of the perforations being thick; there is granular pharyngitis; Eustachian tube dilatable. Applied a cotton pellet; then H. D. was 14-240.

Nov. 6. Tendency to exuberant granulations. *Kali mur.* internally and externally.

Nov. 10. Improved.

Nov. 20. Very much improved; very little pus.

Dec. 3. No discharge; right ear irritable; washing out *débris*, and drying, relieved the irritation. *Kali mur.* internally.

Oct. 6, 1881. Has had no difficulty until present date; now the right ear is inflamed; has slight pain, and the meatus has been moist; *Mt.* not perforated. *Ferrum phos.*

Oct. 7. Tissues very tender; appearances the same. *Ferrum phos.;* in case of pain, *hepar sulph.*

Oct. 10. About the same.

Oct. 18. Right *Mt.* perforated; free discharge of pus. *Kali mur.*

Oct. 25. Adjusted pellet to the perforation. Continued the same remedy.

Oct. 31. Perforation healed. Applied pellet to the perforation on left *Mt.*

I have seen the gentleman frequently up to this time (June, 1883), and he has had no difficulty *since last date*.

Miss L. M., aged sixteen years. *Otitis media suppurativa chronica*. May 12, 1879. Right ear has suppurated occasionally for many years, usually after a head-cold; right meatus filled with pus. On drying the tissues, right *Mt.* is found perforated; the pus secretion is watery and fetid. *Psorinum*.

May 19. Much improved; less pus; perforation the same.

May 26. Suppuration ceased; tissues of the canal dry; perforation healed; pharynx red, granular. *Mercurius dulc.*

May 31. Condition the same; Eustachian tube closed; hearing not improved; H. D. right, 5-240; after inflation, right, 7-240; left, 18-20.

June 7. No special gain for the voice; appearance the same; H. D. right, 24-240; left, 16-20; after inflation, right, 3-20; left, 20-20; Eustachian tubes dilatable. Same remedy.

June 16. Much improved; right, 8-20; left, 20-20.

June 21. Slight improvement; H. D. right, 12-20.

July 2. H. D. 20-20 right and left; right *Mt.* clearer than left.

Mr. J. Q. A. R., aged fifty years. *Polypus. Action of calcarea carb.* March 26, 1872. Had scarlet-fever when a child; right ear involved; almost constant discharge; had a polypus which was removed eight years ago; now worse, H. D. right, minus; left, normal; in the right meatus a broad-based polypus pedicle attached above. Was put on *calcarea carb*

April 6. Ear feels better; less discharge of pus; is sleepy during the day, and sleepless at night, and insists that this is due to the medicine. This remedy was given at intervals until Aug. 13. The discharge nearly ceased, was slightly bloody, and examination from time to time showed a gradual recedence of the growth, and at this date it is nearly gone. *Mt.* clear below, the upper part hidden by remains of the granulating mass; H. D. 20-240; after inflation, 24-240.

Aug. 13. Was put on *mercurius viv.*

Aug. 22. Ear sore to touch. The growth is much smaller, only the base of the growth remained attached to the roof near the upper edge of the tympanum; tissues very sensitive; a probe covered with cotton could not be tolerated for an instant; H. D. after inflation, 26-240.

Aug. 27. Sensation in the ear as if something needed to be removed. Examination showed a small point on the upper edge of *Mt.*, where five days ago there was only a broad base of gray, ragged tissue, very sensitive to touch; H. D. after inflation, 32-240. *Calcarea carb.*

Sept. 14. Better; growth much smaller; still very sore; H. D. the same; discharge excessive.

Sept. 23. Growth larger; has taken cold; hears contact 240. *Calcarea carb.*, locally and internally.

May 16, 1873. Two small granular masses, easily moved, on the upper wall of the canal near *Mt.;* on cleansing the canal, the lower portion of the membrane is seen; tissues not clearly defined; H. D. an inch and a half for a watch heard twenty feet; after inflation, 5-240. *Calcarea iod.*

May 24. Better; appearance much the same.

June 2. A number of small granulated tumors, both anteriorly and posteriorly to the head of the manubrium; H. D. 3-240; after inflation, 5-240.

June 10. Granulations small; less pus; H. D. 6-240. Continue the same remedy.

June 19. Granulations receding, canal more open. Same remedy. This remedy was continued until Nov. 4, when there was a very light discharge: the granulations, which had been excessive, but never amounting to a polypoid form, had decreased very much; and the patient was so well satisfied with the improved condition, that he could not be induced to continue the treatment.

At first I did not accept the statement of the patient as regards the action of the remedy in the matter of causing sleepiness, but was satisfied by repeated experiments that he was peculiarly susceptible to the action of the remedy. The substitution of a *placebo*, and, later, the administration of *calcarea*, reproduced the condition so that he insisted that he could not keep awake during the day and about his work, as a hotel steward, when under the action of the remedy. In this respect the case is unique.

Master Albert A. S., aged thirteen years. *Polypus. Action of alcohol.* May 7, 1881. Two weeks ago noticed a bad smell from his ear. Professor F. S. Bradford, M.D., has treated the case, and now sends him to me. Last summer had ear-ache, and was deaf for a time; no pain recently; dark-colored discharge; polypus in right meatus externus. *Alcohol* instilled into the canal. *Kali phos.* internally.

May 14. Much improved; polypus shrivelled, easily removed with forceps. Continue remedy.

May 21. Tissues dry; instilled fluid *petroleum.* Continue remedy.

May 28. Reports himself all right; tissues are dry, and free from shreds. Continue the remedy.

Sept. 9, 1882. Right ear feels stuffed; again full of white shreds. These are removed by syringing; tissues dry. *Kali phos.*

Sept. 16. Much improved; shreds less; no discharge.

Sept. 30. He has improved. On removing the detritus from the canal, I found fine granulations on *Mt. Kali mur.* This remedy was continued until Oct. 28, when he reported himself all right again. *Boracic acid* was dusted over the granulations.

Nov. 11. Slight moisture.

Nov. 25. Tissues dry and scaling. *Kali mur.* Since that time the patient has been seen, and remains well.

Mr. William E., aged thirty years. *Polypus removed by torsion.* Dec. 7, 1872. No history of early trouble of serious nature. Last August, ear-ache from cold, probably aggravated by habit of picking the canal. Since then, discharge of watery matter, sometimes bloody; H. D. right, 18-20; left, contact; from the right meatus, hard cerumen is removed, then H. D. is 20-20; left meatus filled with a cellular polypus. *Calcarea carb.*

Dec. 13. Same appearance; freer discharge.

Jan. 7 and 31. Same remedy sent by letter, as the patient was a commercial traveller.

Feb. 14. H. D. right, 12-20; after inflation, 16-20; polypus bleeds easily; discharge freer, dark and offensive. *Mercurius.*

Feb. 18. Better; discharge changed to a whitish color; the appearance of the polypus is less cellular.

March 17. Growth filled the meatus nearly to the orifice, and was removed by twisting on the pedicle until torsion removed the entire mass. The bleeding was excessive. *Calcarea carb.*

March 19. The base of the pedicle only remains; H. D. six inches before inflation by Politzer's method, then seven inches. No hemorrhage. *Calcarea carb.*

March 24. Less discharge; somewhat fetid.

March 27. Only base of polypus remains; after cleansing, H. D. 18-240.

March 30. Tissues dry; small prominence where the polypus was attached; H. D. 7-240; after inflation, 10-240.

April 30. No discharge since last date; H. D. right, 18-20; left, 18-240; after inflation, 28-240; the ulcer left by the attachment of the polypus covered with dry secretion. *Silicea.*

May 5. Had the same remedy; H. D. 21-240; after inflation, 30-240.

June 13. H. D. right, 20-20; left, 9-240; after inflation, 1-20; *Mt.* dry, and clear of crust, but dull and thick in appearance.

Miss M. L. B., aged forty years. *Polypus cured by kali bichrom.* Sept. 21, 1878. Has had catarrhal trouble in her ear for nine months; slight pain, then discharge; H. D. 5-240; perforation of membrane in anterior inferior quadrant with polypus, the attachment of which is uncertain; discharge of blood and pus; polypus bleeds easily when touched with the probe; the tissues are extremely sensitive, for which she received *hepar sulph.* The polypus was touched with saturated solution of the *bichromate of potash* and water, under the action of which the growth disappeared in a few weeks.

Feb. 22, 1879. H. D. 8-20; the only complaint is of a subjective sound, like boiling water.

1880. In March she had a slight pain for one day, for which she received *hepar sulph.*

I have heard from the lady through another patient, whom she sent to me, that she remained well since last date.

Miss C., aged about thirty years. *Polypus. Calcarea iod.* Feb. 6, 1874. Has had trouble with her left ear for six years; cause unknown; no particular subjective symptoms, but the annoyance of a constant discharge; left meatus filled with granular polypus. *Calcarea iod.* internally, and the growth was destroyed by *electrolysis.* It returned in about three years, and *calcarea iod.* was given again. The canal was syringed with a solution of *alcohol* and *boracic acid,* and *boracic acid* trituration was used in conjunction with *calcarea iod.* She has had one relapse, due to an acute coryza, which was at once relieved by the same remedies internally, and the same local measures; no return of granulations.

Jane C., aged seven years. *Polypus.* Sept. 16, 1878. She has had a discharge from the left ear for three weeks; had diphtheria and malaria; extremely prostrated; left meatus is filled with pus; inner third of *Mt.* granular, and the tissues undefined. She was put on *baryta mur.* for subjective sounds, on swallowing, and for glandular enlargement.

Oct. 7. Both ears have discharged; the pus is less, and crusts have formed over the ulceration. *Silicea.*

Feb. 5, 1879. The right ear discharges, and the pus is very offensive; on examination, a polypus is discovered in the right meatus; this was very soft, cellular, and easily removed by forceps. *Psorinum* internally, and *alumen ustum* locally.

Feb. 28. Another polypus had grown in the mean time, even larger than the previous one, but of the same nature, and was easily removed; profuse hemorrhage followed.

April 14, 1880. Has remained well since last date, but now has recently relapsed; in the right meatus, pus, and shreds of tissue. *Psorinum* and *silicea.* There were small, scabby sores on the scalp in various spots posteriorly, and the complaint of itching over the whole body.

April 20. Very much better; very little pus; left *Mt.* perforated; mucus was forced through the perforation by inflation. *Psorinum* and *zincum.*

The mother of this child sent another patient to me for treatment, who reported a perfect cure of the girl.

Mr. R. W. R., aged twenty-five years. *Otitis media suppurativa chronica with polypus.* Feb. 8, 1876. For three years and a half has had abscesses in ears, first in the right, then in the left; H. D. right, contact 20; left, 3-240; walls of canal scaly; right *Mt.* irregular and granular; left *Mt.* irregular and dry; tissues of the wall of the pharynx thick and deep red. *Mercurius dulc.*

March 15. Has made weekly visits; H. D. right, 6-240; left, 30-240. Continued under treatment about one year, when H. D. right was 10-240; left, 4-20. The right was still granular. He then passed from under my treatment, and was for four years in the hands of a so-called "regular" specialist.

June 27, 1881. He has had severe pain in the right ear, and returns to me for attention. The pain has been relieved by hot applications. The canal is filled with pus; *Mt.* ulcerated over its whole extent; discharge dark and offensive. *Kali phos.* He was on this remedy till July 8, when he was very much improved; the tissues were clearer, and more defined. Under *kali phos.* he made steady improvement until Oct. 20, when exhaustion with Siegel's otoscope moved the *Mt.* to its full extent; the tympana were dilatable.

Nov. 7. A small polypoid mass was discovered encroaching on the canal from above, soft, easily crushed and removed; very free from hemorrhage on its removal. *Kali chloridum* was given internally, and the hemorrhage was checked by absorbent cotton.

Nov. 26. No discharge, and the tissues are perfectly dry. The patient continued to improve under *kali chloridum* with *mercurius dulc.* for acute conditions through the spring of 1882, since which time he has not returned for treatment.

Mr. J. H. K., aged thirty-five years. *Otitis media suppurativa chronica with polypus.* April 23, 1880. Has had trouble with his right ear since childhood; constant discharge of pus, but no particular subjective symptoms; the meatus is filled with pus. On drying the canal, a polypus is found covering the entire field. The patient was very timid, and objected to its removal. Was put on *kali phos.*, with local applications of the trituration of *sanguinaria nit.* The growth receded for a time, but during the vacation between July and October it became very much larger; and on his return he was put on *kali mur.*, with local applications of *boracic acid*, which checked the growth, and it was easily removed with Sexton's forceps by piecemeal. The *kali mur.* was continued.

Dec. 22. There was no further discharge. The patient was put on *silicea* to heal the ulcer which remained at the seat of the attachment of the pedicle.

The patient remained in good condition until March, when, on contracting a head-cold, granulations were observed in the left side, and there was a slight discharge of pus. He was put on *kali mur.* with *hepar sulph.* at one time for extreme sensitiveness of the tissues.

April 8. The ulceration has healed.

May 11. A relapse again from a head-cold; tissues moist and granular.

June 6. Has been on same remedies to date, and is much better.

July 21. Another relapse; tissues more moist; perforations where the granulations had been.

Oct. 8. Has been only fairly well; tissues moist; perforations the same; head-ache over the right side of the head and temples. *Calcarea sulph.*

Oct. 22. Has continued to improve; the perforations are healed; a simple spot remains on the upper edge of the *Mt.* The tissues remained well until the 28th of June, 1882, when a gland below the auricle, at the ramus of the inferior maxillary, enlarged and suppurated. He was put on *hepar sulph.*, and the abscess opened; there was a free discharge of pus, which entirely relieved the local trouble; the *Mt.* was in the same condition as at the previous visit.

June, 1883. I have recently seen this patient, and he reports that he has remained well since last date.

Case No. 5.—Alonzo Ames, aged forty-one. *Otitis media suppurativa chronica.* Oct. 19. Has been the subject of catarrhal condition of the throat for about three years and a half. Now complains of acute conditions resulting from contracting a cold. Since that exposure, there is an aggravated condition of the naso-pharynx and ear. This is similar to Case No. 2. An acute condition has arisen in connection with previous history of suppurative disease of the middle ear; and there is nothing of special interest to note concerning it, in contrast to No. 2. It is well, however, to examine each case, comparing it with other cases of the same form of disease. In fact, I cannot urge too strongly upon you the practice of frequent examination of the cases which you may have the opportunity to observe. It is only in this way you will be able to recognize, in similar cases, those differences which are very often very important in the selection of the remedy. In this case, the sensitiveness of the tissues suggested the use of *hepar sulph. calc.*, which was given every two hours during the day. *Belladonna*, a solution in water, was given every half-hour in the evening and night, when the symptoms of pain are more marked. This treatment has been the means of great relief. The patient will be directed to present himself again, that we may determine what remedies may be required after the acute symptoms have abated.

Case No. 4.—Garmore Pratt, aged thirty years. *Otitis media suppurativa chronica.* The history of this case is not clear. Some time since, he was attacked by suppurative inflammation of the middle ear, resulting in perforation of the membrana tympani. He was furnished with an artificial membrane, purchased from parties who supply these instruments without any regard to the condition of the patient or the results which follow their use. In this, the artificial membrane produced harm, causing a constant irritation and increased discharge. Coming under our care, he was directed to remove it. Suppuration ceased; and the cotton pellet was substituted, which does not cause any irritation, and gives better results as regards hearing than any artificial membrane possibly can do. In cases of perforations of the membrana tympani, the ossicula are found out of position, and vibrations of the air are not communicated to the stapes in full force. The application of the cotton pellet brings the ossicula more nearly into their normal position, and retains them there, so that the hearing-power is increased to a very great extent in some cases. I have in mind the case of a little girl, in whom the application of the cotton pellet increased the hearing-power for the watch threefold, and for the human voice the gain was much more marked. I will mention, in this connection, a case which interested me very much. The subject was the father of a little patient whom he brought to me for treatment of suppurative disease. He casually mentioned the fact of his own affliction, and showed to me that he had been in the practice of applying a roll of cotton in the ear, in such a way that it rested upon the floor of the canal, and against the drum-head, closing a perforation, and increasing his hearing-power in a very slight degree. I applied a cotton pellet, which the patient removed, finding that he could adjust the roll of cotton so that it served a much better purpose.

Case No. 2. — Horace Trumbull, aged sixty-eight. *Otitis media suppurativa chronica.* Aug. 28. For six weeks has suffered pain. He was affected a few years previously: for the last two years he has been subject to hemorrhage from the nose. The present condition began with itching in the ears: he took a salt-water bath; and there resulted an inflammatory action of the left ear, severe pain, burning, and, finally, a discharge. This is an acute condition occurring in a chronic suppurative case. There has been prevalent among the laity the impression that a discharge from the ear is good, and that, so long as it continues, there is no need of alarm, but, if checked, it will be followed by serious results. This error has been re-enforced, in some cases, by the opinion expressed, and the advice given, by medical men who have not recognized the difference between the effects of disease external to the drum-head, and disease of the middle ear, which has, in some cases, been suppressed by improper local applications. Examination of this case shows the meatus to be filled with pus, removal of which exposes the membrane, more or less disorganized by previous inflammatory action. It is impossible to determine whether the membrana tympani is perforated or not, because of the closure of the Eustachian tube on that side. The patient is not suffering excessive pain, and there is no evidence of general constitutional disturbance. We will give *calcarea phos.*, and dust the surface with a trituration of *boracic acid.*

Sept. 4. The patient was relieved at once; suppuration ceased to-day; the membrana tympani is dry, irregular, and retracted; the Eustachian tube on that side still closed. *Kali mur.*

Mrs. A. H. B. *Mastoid disease.* Dec. 23, 1878. Since six weeks, had suffered pain in the left ear. Poultices were ordered, and continued for five weeks. Now the ear is full of pus, drum-head hidden, Eustachian tube closed; mastoid process swollen; directed to abandon the use of poultices, and simply keep the head wrapped with warm cotton over the affected ear. *Capsicum* in solution every two hours.

Dec. 26. Much better.

Dec. 28. Not so well. Sent to the "New-York Ophthalmic Hospital:" here she was placed in bed, warm cosmoline was dropped into the meatus, and cotton saturated with cosmoline applied over the mastoid process and auricle, secured in its place by a large cap, which covered the whole head. *Capsicum* continued.

Dec. 29. The swelling and tenderness possibly a little less, and the discussion of the question of incision postponed one more day.

Dec. 30. Very much improved; swelling less; tenderness less; pain very much diminished. Continue *capsicum.*

Jan. 15, 1879. Discharged from the "Ophthalmic Hospital" on the 11th, having made a complete recovery from the mastoid disease.

Have seen this lady's physician frequently since that date, and understand from him that the recovery was complete, and the hearing restored perfectly, as good as on the other side.

Cured cases. — Samuel S. *Otitis media suppurativa chronica.* June 15, 1880. Fourteen years ago an insect got in his ear, causing pain for two weeks. By advice, he filled the ear with honey; and, later, a waxy plug came out. Two years ago, he had another attack of pain, and recovered. May 10, he was treated by Dr. Latimer of Brooklyn for a similar attack, and, by his advice, comes to me for treatment. The right meatus is filled by a polypus; walls of meatus covered with pus, on removal of which the tissue is found hard, dry, smooth, and white. I gave him dilute alcohol, to be dropped in the meatus morning and evening; and *calcarea phos.*, to be taken internally.

June 22. A portion of the mass came away this week. The same treatment.

June 29. Tenderness of the canal at the outer third; very slight discharge of pus during the week; polypus shrivelled so that it fills only a very small portion of the canal, and can be easily removed. Continue the dilute alcohol and *calcarea phos.*

July 8. The patient remarked, "I think my ear is all right." In the right ear a perforation of the membrana tympani is seen in its posterior inferior portion. The polypus is shrivelled to a mere point.

This case is quoted to illustrate the effect of alcohol upon granular tissue. I may state, there was no recurrence.

Mr. John O., aged sixteen years. *Mastoid disease.* May 8, 1880. Has been the subject of chronic naso-pharyngeal catarrh for a number of years. Seven weeks ago, a severe cold caused pain in the left ear. A practitioner of the so-called "regular" school performed paracentesis, not being able to relieve the pain with anodynes. A slight discharge followed, but no relief of the pain. The meatus is normal; the inner third reddened; the tissues about the drum-head undefined; hearing, contact-20. *Hepar sulph.* every three hours; *belladonna* to be given in solution every half-hour, in case of pain.

May 9. Pain much relieved—only occasional attacks.

May 10. Much better; itching in the ear; pain only at 3.30 A.M.

May 11. Rather better, but no evidence of pus in the cavity of the tympanum.

May 12. Two heavy perspirations last night; pain set in at 3.45 A.M., and has continued.

May 13. No relief. Consultation with Dr. C. Th. Liebold, who confirmed the diagnosis, and advised removal to the "Ophthalmic Hospital," on which step I had previously insisted.

June 5. On last date, the patient entered the hospital, as suggested, and has remained there until to-day. The mastoid became involved, but, apparently, on its external surface rather than at the antrum, the swelling being above, and on a level with the junction of the auricle with the head, and posteriorly. Upon the determination of pus superiorly, incision was made to the bone, and a free exit for blood secured. The bone was denuded over quite an extent, but no sinus could be detected with the probe. *Silicea* was given, and in two days the denuded bone was covered: the sinus closed from beneath, and healed kindly. The drainage was maintained until the opening closed spontaneously.

June 30. Up to this time the patient has had *calcarea phos.* In general, good condition. The tissues of the drum-head are clearing up, and there is a fair prospect of a good degree of hearing-power.

Sept. 18. Canal and drum-head as clear on the side that has been affected, as on the other; hearing scarcely 10-20 on each side; both drum-heads thickened; the patient the subject of chronic naso-pharyngeal catarrh, which is decidedly ozænic; stricture of the left lachrymal duct. Advised not to think of spending the winter in New-York City.

Cotton pellet. — Miss A. G., aged eighteen years. Dec. 5, 1873. At the age of five years, had suppuration in both ears after scarlet-fever; meatus, right and left, filled with pus; tissues on the region of the drum-head undefined; granulation bleeding easily. The patient had been under old-school treatment, the subject of applications of the nitrate of silver, the granulations being cut out weekly with curved scissors. The patient, under treatment with internal remedies, such as *calcarea iod.*, *kali bich.*, *mercurius*, *silicea*, and *tellurium*, and local application of *alumen ustum*, improved so, that, by January, 1876, the granulations had disappeared, and a large perforation was distinguished right and left. The patient continued to improve till the fall of 1880, when she was put upon *kali mur.*, under which remedy she has continued, with an occasional prescription of *mercurius*, in some form, or *hepar sulph. calc.*, for acute symptoms. Since February, 1880, she has had the cotton pellet applied on both sides, whenever occasions have occurred on which it was especially desirable that she should be able to hear more promptly than under ordinary circumstances. On the succeeding days, the pellets would be removed, on account of the suppuration. Since 1881 she has been able to wear the cotton pellet more or less constantly in the left ear, the right ear suppurating.

I may add, that since October, 1881, she has improved decidedly by the local use of boracic acid trituration. I now apply the pellet in each ear.

Miss A. R., aged nine years and a half. May 15, 1880. When about a year old, had discharge from the left ear during dentition. Was treated, in 1875, for ulceration of the left ear, which healed under *tellurium*. At the present date, there remains only the falciform border of the drum-head; the remains of the manubrium malleus are drawn inwards to the tympanic wall; hears only 12-240; on adjusting the cotton pellet, hears 28-240.

Feb. 7, 1882. The cotton pellet remains in position; the

patient hears 46-240. A crust of cerumen, mixed with epidermis, having accumulated, the pellet was removed; hears 7-240. The pellet replaced, hears 28-240. This case is cited, simply to show the decided improvement in hearing; though I believe that it is true, that the presence of the cotton pellet acts as a prophylactic measure, because the mucous membrane of the tympanic cavity is not exposed to the direct effects of the cold air.

Case No. 3.—Christy Powers, aged 12 years. *Otitis media suppurativa chronica.* Oct. 12, 1882. Four years ago, this lad was attacked by measles, which left its sequel in the form of a discharge from the left ear; subsequently, a slight flow from the right ear, followed by deafness. From his own statement, the disease is not associated with pain. The case illustrates the results of exanthemata in the form of measles. The disease began in the naso-pharynx, extended along the Eustachian tube, and involved the middle ear. The membrana tympani yielded to the inflammatory action. These discharges gave a basis upon which extensive changes have been built up,—changes in the epithelium; later in the deeper tissues, finally resulting in extensive perforations. The right meatus is now filled with granulations, which are very extensive, covering the posterior wall, as well as the membrana tympani, so that the relations of the parts cannot be determined. The left ear is in a similar state, but the ulceration is what may be termed indolent. The membrana tympani is perforated, but there is no evidence of exuberant granulations. The treatment has both been mechanical and medicinal. The first was to overcome the fetor, which marks the suppurative process in many of these cases. To accomplish this, nothing is better than the boracic acid trituration, which may be either fifty per cent or seventy-five per cent trituration with sugar of milk. Whether there be any truth in the theory of micro-organisms, as related to suppurative process, has been seriously questioned; but one of our famous German aurists has demonstrated that micro-organisms do exist in the secretions as long as the suppurative process is active, and that as soon as the micro-organisms are destroyed, by the use of antiseptic agents, the secretions cease to be suppurative, and most cases hasten to a cure. Mechanical treatment in this case was, to inflate the middle ear, through the Eustachian tube, forcing the drum-head outward towards its normal position, from which it had been pressed inward by atmospheric pressure. Inflation was performed by the use of the Eustachian catheter, through which air is forced, dilating the tube, and throwing the drum-head outward.

Oct. 26. This case returns to us to-day for examination, to note the improvement. The treatment which was directed on the former day has been continued. The granulations of the right ear are clearing up, and suppuration is much less marked. In this case, I do not think that the change is due to the use of the boracic acid alone, nor do I think that the internal remedies have done the work exclusively: both means have worked hand in hand. In cases similar to this, I have often followed the suggestion given by some foreign writer, — whose name at this moment I cannot recall, — of using a mixture of alcohol in warm water for syringing the ear in suppurative cases. I have had adjusted for my use an arrangement, which I have never seen suggested by another. It consists of a suitable tip, fitted to an ordinary ear-syringe, such as is fitted to the ear-syringe for a separable nose-syringe. On this tip I adjust the ordinary hard-rubber Eustachian catheter. This enables me to turn the handle to either the right or left side; and, while the meatus is well illuminated, the beak of the instrument can be passed carefully in, either with or without the speculum, thus bringing the instrument to bear in any direction, and upon all parts of the illuminated surface. Afterwards, the tissues must be perfectly dried, then the boracic acid dusted on, as before suggested. In some cases, it is desirable to avoid the use of the snare for the removal of large granulation masses or fibrous polypi. For this purpose, I have found boracic acid less irritating than bichromate of potash, as suggested by Dr. W. P. Fowler of Rochester. I always have prepared a saturated solution of boracic acid in water. I take equal portions of this and common alcohol, and direct the parent, or person having charge of the patient, to drop a few drops, say five, in the ear night and morning, closing the canal with cotton. After a few applications, the growth is checked; and the application of the dry boracic acid, at the office-visits, is evidently more effective. In a short time, one is able to remove the remainder of the mass with ordinary forceps, without disturbing the equanimity of the little patient.

March, 1883. This patient has made rapid progress under *kali mur.* and the local treatment. The granulations have entirely disappeared from the right ear. The right membrana tympani is dry, adherent, with deep folds anteriorly and posteriorly to the manubrium, and perforated in the anterior inferior quadrant. The left membrana tympani has cleared up, is perforated in the anterior inferior quadrant; and the cavity of the tympanum is nearly free from purulent products under the action of *silicea.*

Mr. G. C., aged sixty-two years. *Mastoid disease.* Jan. 11, 1873. He stated, that, thirty-two years ago, he had an "awful" pain in the left ear; after some time heard as well as ever. Last October took cold; the left ear pained, finally broke, then healed, and broke a second time; hearing with right ear, 12-240; left, 2-240; the right meatus narrow, the left filled with fetid pus; mastoid process sensitive, red, swollen; whole side of the face aching, a dull ache, at night worse. *Capsicum* every two hours; *mercurius* at night, every half-hour, when the pain is aggravated.

Jan. 22. A large tumor behind the left auricle, displacing the auricle altogether; less pain; the meatus full of pus, but no escape from the cavity of the tympanum. *Capsicum* in solution every two or three hours during the day.

Jan. 23. More pain last night, and more discharge; the same appearance. Continue *capsicum* and *mercurius* as before.

Jan. 24. A better night; only discharge from the meatus; mastoid process less swollen and sensitive. Continue the same remedies.

Jan. 27. As the condition was not materially changed, decided to make an incision in the mastoid, cutting down to the bone: the knife passed readily into the antrum; and a free discharge of pus followed, dark, bloody, and very offensive. *Hepar sulph.* The free discharge of pus from the mastoid relieved the pain; discharge from the meatus freer. *Capsicum* and *silicea.*

Feb. 1. Less discharge from mastoid process and meatus. Continue the same remedies.

Feb. 3. Only blood flows from the mastoid process, and pus from the canal.

Feb. 5. Not as well; discharge from the meatus less; discharge from the mastoid process increased. *Capsicum.*

Feb. 7. More pain at night; free discharge of blood and lymph from the sinus in the mastoid process, but thick pus from the meatus. *Capsicum* through the day, and *mercurius* at night.

Feb. 10. Much better; no pus in meatus; sensation as of movement in the ear when chewing; watery discharge from the mastoid process; pus in the canal.

Feb. 14. Much better; only one slight attack of pain; tissues of the mastoid process much freer; watery discharge from both meatus and mastoid process.

April 14. The same conditions have continued up to this day, with slight but decided improvement. On this date I used the method, suggested by Dr. James Hinton, of forcing some fluid, by way of the meatus, out through the sinus of the mastoid. For this purpose I covered the tip of the ear-syringe with a section of a rubber tubing, closing the meatus completely, filled the ear-syringe with warm fluid cosmoline, and with steady pressure entered the syringe into the meatus and tympanum. Pus and blood were driven out through the sinus of the mastoid process into the pharynx by the Eustachian tube. A second injection resulted in bringing the fluid petroleum through nearly free from pus. *Capsicum.*

April 23. Much better; Hinton's method continued. *Silicea.*

May 3. Hinton's method has been used thoroughly; little pus in the canal; perforation of the drum-head is well defined, and the passage of the fluid petroleum through the sinus is much more free and easy than previously. This method was continued until July 18, until both from the canal and opening in the mastoid the discharge was nearly clear and lymph-like fluid. The patient then complained of severe neuralgia, and disturbance of the stomach and abdominal viscera, so that he became very intolerant of the treatment for the ear; and from that time I lost track of him.

I learned, after the first notice of the case, that the patient was treated, immediately after the first pain in the ear, by a so-called "botanic" physician, who ordered large fomentations of herbs applied over the whole side of the head; and these were changed from hour to hour, and continued from day to day, until the suppuration was established to the degree which existed at the time of his visit to me. From the statement of his daughter, I am satisfied that the excessive heat and moisture tended to increase the suppuration, while it did not relieve the pain.

SUMMARY OF REMEDIES.

Aconite. — In acute suppuration ot the middle ear, or for acute symptoms arising in chronic disease. (See External Ear.)

Agaricus muscarius. — For spasm of pharyngeal and tympanic muscles; twitching and rattling or fluttering in the tympanic cavity (R); jumping of tensor tympani, with sound as of a leather-covered metal valve; rolling, twitching in the tympanum; creaking in both ears on empty swallowing; at every attempt to swallow, a creaking sound in both ears, as of a wooden screw.

Aurum metallicum. — This is indicated in consequence of suppurative inflammation of the middle ear. The periosteum of the temporal bone being affected, the porous diploe yield; and finally the entire surfaces, external and internal, are denuded. The subjective symptoms, so far as the ear is concerned, are decidedly negative; but the general ones make the choice between this remedy and fluoric acid, nitric acid, or silicea, easy. Pain, like a bruise, or as if pulled, worse at night by uncovering and at rest; better by motion, by washing, and while sensitive to cold; yet relieved by going into the open air, even in bad weather. The tissues of the *Mc.* are bathed by a fetid pus, the odor being characteristic of necrosed bone. The *Mt.* is usually perforated, the ossicula more or less disintegrated and thrown off. Often the osseous meatus is denuded, sinuses connecting the canal with the fistulous openings upon the external surface of the mastoid process.

Baryta muriatica. — Baryta is one of our most valuable remedies, both in suppurative and non-suppurative inflammation of the middle ear. After frequent failures with the carbonate, we find that the symptoms, as given in Allen's "Encylopædia," vol. ii., Baryta carbonica, — symptom 122, "hardness of hearing;" 123, "severe buzzing in the ears;" 124, "cracking in both ears when swallowing;" 125, "cracking in one ear when swallowing, as if it were breaking;" 126, "cracking in one ear . . . when swallowing, sneezing, etc.;" 129, "a reverberation in the ear on blowing the nose violently," — have been repeatedly relieved by the muriate. What is the significance of these symptoms? Dr. Hinton, in his work entitled, "Questions of Aural Surgery," writes, "Mr. Yule confirms the statements of Drs. Jago and Rumbold respecting the effect of an abnormally open Eustachian tube in intensifying any sounds produced in the patient's own throat; and I had the opportunity of demonstrating that the cause is rightly assigned, by introducing into his tube a vulcanite Eustachian catheter, in the curve of which an orifice was cut, so as to establish a continuous passage from the throat to within the tube, when the very same effects that result from his own muscular action were produced." Those specially interested in the details of this study are referred to the above-mentioned work.[1] These

[1] James Hinton, The Questions of Aural Surgery, p. 102. London, 1874.

symptoms are, therefore, due to interference with the action of those muscles which open and close the Eustachian tube. We shall have occasion to contrast an abnormally open tube with one closed by morbid conditions.

Belladonna. — In acute inflammation of the middle ear, or when acute symptoms arise in chronic disease. (See External Ear.)

Calcarea carbonica applies to the same class of patients as in general diseases, — the fat, rapidly growing, large-headed, soft-boned children, or adults who in youth were vigorous, but now fail from low power of assimilation, great weakness, dejection, sensitiveness to cold, damp air. The pains about the head are pressing or pulsating, often semilateral; coldness of the head; sweat on the head evenings. The pain in the head is also beating, with knocking, buzzing, and roaring. Detonation in the ears. Meatus filled with whitish, pappy, fetid pus, or viscid discharge. Membrana tympani perforated, and often the edges are covered with granulations which extend to the walls of the meatus: occasionally these enlarge, and form polypi, usually of the mucous, or cellular, variety. We have found these exuberant granulations to yield more promptly to the *calc. iod.* than to the *calc. carb.* If the growth is large, and cause pressure by checking the escape of pus from the tympanum, it should be removed by mechanical means; but this does not prevent its rapid renewal. *Silicea* should follow the *calcarea* after the ulceration assumes an indolent type.

Capsicum. — For chronic suppuration, in adults especially. The pains in and around the ear are acute, shooting, pressing, with bursting head-ache; great thirst, with chilliness and shivering. In February, 1872, Dr. T. F. Allen called our attention to this special symptom: "On the petrous bone, behind the ear, a swelling painful to the touch." In April, 1873, we published, in the Ophthalmic Hospital Reports, for the New-York "Journal of Homœopathy," cases showing its value. Ten years have added many cases to our list of cures: the typical ones are those in which acute symptoms occur in chronic cases; the mastoid cells become involved, and their dense structure yields slowly; hence the danger of cerebral trouble, as the diploe of the temporal bone above threatens to give way before the petrous portion below and behind. In children, the mastoid cells are large, and, with their walls, break down with comparative ease. *Hepar sulph. calc.* hastens the relief, when the case has advanced far before capsicum is used. In some cases, the swelling behind the auricle has been very great, turning it almost to a right angle with the side of the head, the meatus being closed almost entirely; the pus is yellow, flowing quite freely, and not especially offensive. In every case the *Mt.* was perforated.

Carbo animalis and carbo vegetabilis are found in our repertories as curative for otorrhœa; and one writer states, "suppuration of the internal ear, and discharge from the same," meaning middle ear. Thus far, all the cured cases seen at the clinic and in private practice were those where the lesion was external to the membrana tympani. A review of the year-books confirms our view, that the action of the remedy is curative in chronic non-suppurative inflammation of the middle ear,—the so-called proliferous form. Objective symptoms: membrana tympani retracted, opaque in most cases, tympanum dry, the Eustachian tubes easily dilated, the pharynx granular. The subjective symptoms suggest a dry state of tissues, itching and tickling in the ear (similar to the throat-symptoms), with cracking on moving the jaw, an inclination to swallow, which relieves the ear.

Causticum.—There is an absence of objective symptoms in cases cured by causticum, but the subjective ones are very pronounced: "Crawling, as from an insect;" "Itching in the ear, beginning in the throat, along the Eustachian tube." (Also under *nux. vom.*) The symptoms under the internal ear are noteworthy.

Chamomilla. In acute disease, with the symptoms noted under otitis externa.

China, or cinchona.—Indicated in suppurative inflammation. After repeated trial of the various remedies given in our repertories for hemorrhage from the ear, such as *cicuta, lachesis, mercurius, pulsatilla*, and failing of satisfactory results, we gave *china* to a little girl at the clinic of the New-York Ophthalmic Hospital. The following were the objective symptoms: meatus full of sanguineus, sanious discharge; the tissues infiltrated, and suggestive of deep-seated disease; *Mt.* not seen: the patient was very anæmic; and the cinchona was given upon general, rather than local, indications To our surprise and delight, the case changed its features at once: the meatus became more open, the membrana tympani defined, the perforation clearly so, the flow of blood ceased, and the pus became more laudable. Since that date, we have used cinchona in every case of similar nature, with a prompt response. The condition of tissue is not that of exuberant granulation, but indolent ulceration, with passive hemorrhage. In one case, there was mastoid disease of long standing; a sinus opened upon carious bone. The action of cinchona upon the middle ear, in non-suppurative inflammation, is a subject upon which we are not prepared to speak as we desire. That it causes hyperæmia of the tympanum has been shown by Hammond and others. Roosa[1] reports a case of otitis, which illustrates the action of quinine on the external ear as well as the labyrinth. We have a case under treatment, where the otitis was aggravated, if not caused, by quinine, the acute symptoms recurring every seventh day. In two cured cases,—one in our own practice, the second reported by a colleague,—the lesions were not of the external or middle ear, but clearly of the internal. Of these cases we will speak when considering the therapeutics of the internal ear.

Elaps corallinus.—Indicated in the chronic suppurative form of disease, complicated with naso-pharyngeal catarrh; the posterior wall of the pharynx covered with crusts, or mucous membrane fissured; nasal mucous membrane in same condition; external meatus full of offensive yellowish-green discharge, which stains the linen green; membrana tympani

[1] Roosa on the Ear, p. 169. New-York City, 1885.

usually perforated. Subjective symptoms: congestive lancinating frontal and occipital head-ache, aggravated by motion and stooping. This remedy is of great value in the naso-pharyngeal catarrh, which complicates aural disease in children. The patients are compelled to sleep with the mouth open, on account of the obstruction of the nares; hence the term snuffles, used by mothers and nurses.

Ferrum phosphoricum. — Schüssler claims that this salt controls the beginning of disease. "Whilst iron restores to their normal condition the blood-vessels, enlarged by disease, it heals the irritation-hyperæmia, which is the cause of the first stage of all inflammations: . . . fresh, non-suppurating wounds are quickly healed by this remedy." This remedy has been called, "tissue aconite." One characteristic symptom may guide to its use, — beating in the ear and head: the pulse can be counted in the ear, one patient remarked. The symptomotology of ferrum met. must be studied for confirmations.

Gelsemium. — While this remedy may be more frequently needed in acute disease of the middle ear, it may be specially effective in mastoid disease, or acute necrosis, complicating acute suppuration. (See Otitis Externa.)

Graphites. — The relation of this remedy to the nutrition of the skin holds good in dry conditions of the mucous membrane: indeed, we may infer very much of the condition of the tympanum from study of the dermoid layer of the external auditory canal. Hence the condition is that of sclerosis or proliferous inflammation. The membrana tympani may be opaque and thick, or transparent and very thin, adherent to ossicula or promontory, or perhaps mobile; Eustachian tube dilatable, but hearing not improved by inflation. There is one subjective symptom which is characteristic, — "hearing improved in a noise." Writers upon aural surgery differ widely regarding this matter. Von Troltsch treats the matter very lightly, as if the patient's statements were faulty. Kramer and Wilde accept the fact, but fail to account for it. Dr. Peter Allen,[1] Aural Surgeon to St. Mary's Hospital,

[1] Lectures on Aural Catarrh, p. 29. London, 1870.

London, England, gives an explanation which is satisfactory, one which we have been able to confirm. Space will not suffice to go into the details of the physiological action of the muscles of the tympanum: we will simply state, that any change in the tympanum which either presses the stapes into the fenestra ovalis, or draws it out beyond normal relations, changes the tension of the serous fluids of the internal ear; and this causes subjective sounds. Such changes do occur in the dry, atrophied condition we are considering, and the atrophied muscles cannot sustain the ossicula in their normal relations under ordinary sound-waves: when, however, the patient is subjected to extraordinary continuous sound-waves, a reflex action stimulates the muscles, the sound ceases, and hearing improves. This can be done by electricity, as we have shown in our article read before the Homœopathic Medical Society of the State of New York, 1875. That which is done by the heavy sound-waves, or by electricity, is accomplished by graphites in a gradual manner: the function of the tympanic muscles is restored when the nutrition is re-established.

Hepar sulphuris calcarea.—In the suppurative forms; membrana tympani perforated; ulceration angry; discharge small amount, sour, fetid odor; the tissues very sensitive, often covered with white shreds, which cling to the ulcer. Subjective symptoms: soreness in small spots about the ear; itching; patient worse at night and by cold air.

Hydrastis canadensis stands first among remedies for *muco-purulent* discharge from the middle ear. We should expect this from its action on mucous membranes elsewhere. In purulent inflammation of the middle ear, with thick, tenacious discharge, more mucus than pus, this remedy is invaluable. Compared with kali bichromicum, there is less tendency to granulations about the perforation of the membrana tympani, less cracking and crusting of mucous surfaces.

Iodine.—In chronic non-suppurative disease. Curative in atrophy of mucous membrane, probably by stimulating glandular elements of structure. (See *graphites.*)

Kali bichromicum. — In chronic suppuration; the membrana tympani perforated; the cicatrization of the edges of the perforation complete; the tissues have an appearance as if changed to mucous membrane, and the secretion is often more mucus than pus; the discharge yellow, thick, tenacious, so that it may be drawn through the perforation in strings. The subjective symptoms are, lancinations, sticking sensations, that the patients are not able to locate with any degree of positiveness.

Kali hydriodicum. — The iodide of potash is a valuable remedy in catarrhal, suppurative, or serous disease, if the diagnosis of specific disease is clearly made. In attenuation, it is a parallel of acute and chronic naso-pharyngeal catarrh.

Kali muriaticum (Schüssler). — One of the most effective remedies we have ever used for chronic catarrhal inflammation of the middle ear, specially of the form designated "proliferous." The subjective symptoms, as learned from clinical use, are not numerous: a stuffy sensation in the recent cases, subjective sounds, and deafness, are very marked. The objective symptoms are, the naso-pharyngeal obstruction, the effort to clear the fauces, the granular pharyngitis, pharyngeal tonsil, closed Eustachian tube, retracted *Mt.*, and atrophied walls of the external meatus. In chronic suppuration, it reduces proliferation, checks granulation, hastens repair, probably by action on mucous membrane (compare *silicea*), and gives an increased power to withstand exposure. Its action seems more decided on the right Et. (Compare *merc. dent.*)

Kali phosphoricum (Schüssler). — For suppurative disease, specially chronic form, Schüssler writes thus: "Potassium phosphate cures the following diseased conditions: septic, scorbutic bleedings, mortification, encephaloid cancer, gangrenous croup, phagedenic chancre, putrid-smelling diarrhœa, adynamic typhoid conditions, etc." From the foregoing indications, we were led to use it in ulceration of the membrana tympani, with or without perforation, in suppuration of middle ear, the pus being watery, dirty, brownish, very fetid, the ulceration angry, bleeding easily, and showing little tendency to granulate, or secrete laudable pus. It is specially valuable in old people.

Kali sulphuricum (Schüssler). — For catarrhal disease or suppuration, if the discharge be muco-purulent rather than purulent. The guiding symptom is the color of the secretion, which is yellow and sticky, tenacious. It has an action on the Eustachian tube similar to the kali muriaticum.

Lachesis. — This remedy has been curative in some cases of the proliferous form, the objective symptoms being similar to those given under graphites: the subjective symptoms of left-sided deafness, and aggravation after sleeping, led to its selection in lesions after typhoid or typhus fever and diphtheria.

Lycopodium. — For scrofulous subjects, those who suffer from moist eruptions, abdominal troubles, sequela of scarlet-fever, etc. Hence used in the suppurative form: the membrana tympani is often destroyed, with the exception of a small margin, the ulceration showing no disposition to heal, but not sensitive to touch, as under hepar; the pus is offensive, not profuse, and inclined to crusts if not frequently cleansed.

Magnesium phosphate (Schüssler). — The action claimed for this remedy is upon the muscles, nerves, brain, bones, and teeth. It is very effective in neuralgia. Dr. William E. Rounds (my colleague in the Ophthalmic Hospital) suggested its use in the proliferous form of middle-ear disease. I believe it supplements the kali mur.

Mercurius biniodide. — For similar conditions to those for which the metal has been given, but its action on the left Eustachian tube is the special indication for administration.

Mercurius dulcis. — In the "Homœopathic Times," April, 1875, we gave the reason for employing this remedy in chronic catarrhal inflammation of the middle ear. Time has confirmed the anticipations raised by its first trials. The objective symptoms are those of this form of inflammation, — the membrana tympani retracted, thickened, and immovable by inflation; a granular or hypertrophied condition of the pharyngeal mucous membrane. The subjective ones are those of a benumbed, dull feeling between the throat and ear, a pressure in the ear from without; with these, the subjective symptoms of irritation of the auditory nerve from pressure or tension on the stapes and fenestra rotunda, humming, roaring, and singing: these fade as the audition increases. In many cases, the more intelligent patients can give the moment when the air passes, for the first time in months or years, from the throat to the ear. Of this form of disease, Toynbee writes thus: "This affection is less prevalent in the young than in adults, in whom it is liable to occur after repeated attacks of cold, whether with pain or not. It is, however, most frequent in persons advancing in life, and may, in fact, be considered as *the* disease which causes *deafness in advancing years.* The generally received opinion, that in this kind of deafness the nervous system is at fault, is manifestly incorrect, as proved by the symptoms, and by the mode of relief found beneficial." It is to this condition that mercurius dulcis is proven to be homœopathic.

Mercurius protoiodide. — The objective indications for the iodides of mercury are, the thickness of the tissues of the fauces, the lobulated tonsils, with deep intra-spaces. The protoiodide acts specially on the right side.

Mercurius vivus. — In chronic suppuration, with enlarged and sensitive cervical glands, resulting from sudden colds. The membrana tympani is not broken down to a very great degree, and often repairs very promptly, even in cases where the perforation has existed for months: the discharge is pus, fetid, whitish, or mingled with blood. Subjective symptoms: tearing pains with the discharge, these aggravated by any cold which checks the free flow of pus; also worse by lying on the ear in bed.

Mezereum is valuable in some cases of sensitiveness of the mucous membrane of the middle ear. "The ears feel as if too open, as if the tympanum were exposed to the cold air, and it blew into the ear." This has been repeatedly confirmed.

Nitric acid. — In ulceration, caries of the ossicula or mastoid process, the remote results of syphilis or abuse of mercury. Subjective symptoms: shooting pains; sensitiveness of the bones; aggravated by every change of temperature, also at night, on waking, or rising from a seat, and by touch (see Aurum); better while riding in a carriage. Fluoric acid has inclination to uncover and to wash with cold water (see *silicea*).

Nux vomica. — In irritation of the mucous membrane of the middle ear; "itching in the Eustachian tube, provoking swallowing." Similar to causticum symptom.

Phosphorus corresponds to a dry condition of the tympanum. One objective symptom, deafness, is interesting in this respect, that the failure is especially for the human voice: noises and musical tones are recognized much more readily than the modulations of voice. It is undoubtedly true that we must look for an explanation of this fact in the functions of the tympanum, and hope that a more complete knowledge of its physiology will reveal the reason.

Phytolacca decandra. — Occasionally indicated in acute inflammation of the middle ear: the symptoms of the pharynx and Eustachian tube guide in the selection. Shooting pains in both ears when swallowing; worse on right side; Eustachian tube obstructed; fauces dark, bluish-red; tonsils large, bluish, ulcerated; throat feels rough, hot, dry, with burning smarting. Valuable in diphtheria and malignant form of scarlet-fever.

Psorinum. — A remedy closely allied to sulphur. Bell[1] writes, "Whether derived from purest gold or purest filth, our gratitude for its excellent services forbids us to inquire or care." In chronic suppuration, where the symptoms remain unchanged after sulphur, the ulcers scab over rapidly; the pus *very fetid*, with the ulceration of the membrana tympani; scabby ulcers on the vertex and behind the ears. Subjective symptoms: excessive itching in the ears, so that children can hardly be kept from picking or boring in the meatus.

[1] Bell, on Diarrhœa and Dysentery. New-York City, 1869.

Pulsatilla. — For acute catarrhal inflammation, or chronic suppuration, when the discharge is a bland muco-purulent secretion. Fever without thirst, relief of pains in the open air, and a peevish, changeable, timid disposition, indicating the nervous depression, are guiding symptoms.

Quinine. — The muriate of quinine is the remedy for the conditions sometimes noticed in acute suppuration, when the exudation of lymph and pus suddenly ceases, and the deeper tissues become involved, thus threatening death by acute necrosis or cerebral lesion. The case which led me to the use of the drug was unique. A young physician, who had had chronic catarrhal otitis for many years, was attacked with acute otitis media: the membrana tympani was very thick and unyielding; upon incising it freely, the pain was relieved; a free discharge followed; twenty-four hours later, the flow ceased quite suddenly, and the case became critical. Sickness prevented my attention to the patient at his house. Dr. Rounds saw him, at my request: he called Dr. Liebold for counsel, who advised the muriate of quinine; it was given, with prompt and permanent results. Since then, we have used it in all similar cases, as well as in any with marked periodicity either of pain or secretion.

Silicea. — In chronic suppuration; ulceration in cachectic subjects, or those who have been dosed with mercury; in caries or necrosis. Objective symptoms: membrana tympani perforated and irregular; secretion of pus scanty; ulcers deep, and covered with scabs unless frequently cleansed. The theoretical relation of silicea to the periosteum would lead us to expect curative action when the middle layer of the membrana tympani (*substantia propria*) is involved, and clinical results do not disappoint us. It has been our view, that more repairs of the membrane occur under the use of this remedy, in chronic disease, than under any other single remedy.

Sulphur. — The indications for this remedy must be sought in general rather than in the special objective and subjective ones, as they are meagre compared with the last-mentioned remedy as well as others. Itching in the ears, drawing or shooting pains in the ears; discharge of pus, stinking, with crusts.

Teucrium marum verum. — This is the only remedy under which we are able to say we have seen mucus accumulations disappear from the tympanum. In one case, the Politzer's method was used in connection with remedies: the accumulation persisted till teucrium was used. "One swallow does not make a summer." We are cautious in attributing to remedies that which belongs to mechanical means. The symptoms, as given by Dr. Dudgeon, certainly suggest the above condition: "Sometimes fine ringing in right ear, on blowing the nose; a peculiar chirping sound, as if air passed itself through mucus; thereafter the ear remained some time stopped up, and opened up again with a dull sound."

Tellurium. — Curative in chronic suppuration, when the symptoms correspond to the following: a watery fluid, smelling like fish-pickle, which excoriates the meatus and the skin wherever it flows. After the suppuration has ceased, the membrane has been found cicatrized and corrugated, but not thickened.

Thuya oc. — The special indication for this remedy is the discharge "smelling like putrid meat." Clinically, it has cured granulations in the meatus similar to condylomata.

PART THIRD.

DISEASES OF THE INTERNAL EAR.

ELEVENTH LECTURE.

INFLAMMATION OF THE INTERNAL EAR.

Diseases of Internal Ear. — The consideration of this subject will serve to illustrate what I said at the outset, of the two ways in which knowledge of the diseases of the ear is obtained.

The literature upon *otitis interna* is made, upon the one hand, of clinical cases, in which the history of the patient shows more or less evidence of disease which can be referred to the tissues of the labyrinthine acoustic nerve in its branches or origin, or both. Again, a second class of cases are those in which *post-mortem* examinations have revealed lesions of the sound-perceptive apparatus, independent of, or secondary to, disease of the tympanum. Unfortunately, however, these cases have not been identical, with a very few exceptions: hence the pathological results cannot be explained by the clinical history; nor, on the other hand, has history of the life-experience been explained by the *post-mortem* examination. Enough, however, is understood to warrant our assertion, that there is a very close parallelism between diseases of the eye and the ear, both as regards the conducting and the perceiving portions of the apparatus. Formerly, diseases of the nervous portion of the auditory mechanism were as obscure as were the diseases of the fundus of the eye previous to the invention of the ophthalmoscope; and the term "nervous deafness" had as wide and as vague a meaning as the term "amaurosis" formerly had. Careful clinical observation and pathological research have cleared this field of study of its uncertainties; and, while there is much to be done, time will give us as clear and satisfactory a classification of diseases of the internal ear as we now have of the middle ear. There is no reason in the

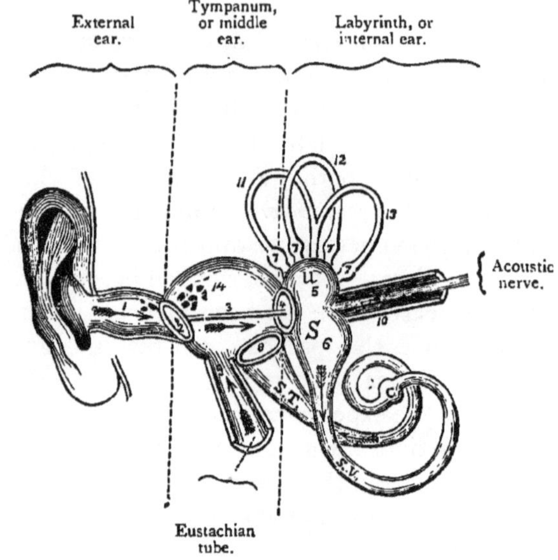

A DIAGRAM DESIGNED TO ILLUSTRATE THE PHYSIOLOGY OF THE LABYRINTH. (PROFESSOR A. L. RANNEY.)

1. External auditory canal 2. The membrana tympani. 3. The tympanic cavity with its chain of bones connecting 2 with 4 4. The fenestra ovalis. 5. The utricle, communicating with the semicircular canals (11, 12, and 13). 6. The saccule, communicating with the scala vestibuli of the cochlea (*S.V.*). 7. The ampullæ. 8. The fenestra rotunda, opening from the scala tympani (*S.T.*) into the tympanum (3). 9. The Eustachian tube, allowing the entrance of air from the pharynx into the tympanic cavity. 10. The internal auditory canal, transmitting the acoustic nerve. 11, 12, 13. The semicircular canals. 14. The opening of the mastoid cells into the tympanic cavity (3), and the external auditory canal (1). *S.V.* Scala vestibuli of the cochlea. *S.T.* Scala tympani of the cochlea. *C.* Cupola. (From Roosa's Treatise.)

nature of things why we should not have hyperæmia, inflammatory action of the higher grades, serous exudations over more or less of the auditory tract, hemorrhagic exudations, or even suppurative processes. These may be primary, arising from no traceable causes,—hence as truly idiopathic as similar diseases in other portions of the body. That they are also secondary to inflammations of the cavity of the tympanum, or to the cranial cavity, we have every reason to believe, from evidences afforded by *post-mortem* examinations. Indeed, inflammation may result in exudation which may be quite local, and involve only a portion of the terminal mechanism of the auditory nerve in the labyrinth; and the results of this exudative disease may be permanent, or may entirely disappear. The earliest observations on this subject, tending to the recognition and classification of the diseases of the labyrinth, were made by Menière, in 1860. This gentleman called attention to certain cases characterized by symptoms which had been considered cerebral in their origin, such as sudden deafness, dizziness, nausea and vomiting, and loss of equilibrium. These cases led to his endeavor to confirm the following propositions:—

1. "An auditory apparatus, hitherto perfectly normal, may become suddenly the seat of functional disturbances, consisting in noises of a variable nature, continuous or intermittent, and which may be accompanied, sooner or later, by a diminution in hearing."

2. "These functional troubles, having their seat in the internal auditory apparatus, may give rise to symptoms which have been considered cerebral, such as intense vertigo, uncertainty of gait, turnings to the right or left, and falling; and they may be attended with nausea, vomiting, and syncope."

3. "These accidents, which are of intermittent type, are at last followed by deafness, gradually growing worse; and often the hearing is at last suddenly and totally lost."

4. "All this tends to confirm the belief that the lesion, which is the cause of these functional troubles, is in the semicircular canals."

Burnett,[1] in commenting on these propositions, calls attention to the fact, that some of the symptoms on which Menière laid stress are not always present in disease of the internal ear; also that some of these symptoms may be caused by irritation of the external ear, as well as lesion of the tym-

[1] Treatise on the Ear. Charles H. Burnett, M.D. Philadelphia, Penn., 1877.

panum: still, to the French author belongs the honor of initiating a closer analysis of these conditions; although exceptions have been taken to some of his conclusions.

Subsequent study by other observers has led to a still closer differential diagnosis between diseases affecting the labyrinth, and those affecting the labyrinth, brain, and spinal cord, or both; and it is now well established, that diseases involving the labyrinth alone were mistaken for diseases of the brain. Before speaking of actual inflammation, I wish to state, as summarily as possible, the points of the condition now recognized as auditory vertigo, or labyrinthine vertigo, independent of organic disease. The mention of a case which I saw in consultation with a medical gentleman of this city will illustrate this matter. A gentleman, who had usually enjoyed good health, had overtaxed himself in his daily occupation, — that of a court-stenographer, — and hence he was below standard as regarded his general health. He was suddenly alarmed by an attack while in the court-room, having many symptoms suggesting some lesion of the brain. He became dizzy, nauseated, unable to sustain the centre of gravity, and would have fallen save for assistance. He became partially deaf, and, withal, very confused mentally. At no time was the pulse above the normal, nor was there any evidence of constitutional disturbance, such as a rise in temperature, etc. On being conveyed to his home, this attack passed without paralysis of any function; and the hearing came slowly back to its usual standard. He was obliged to keep a recumbent position; as any effort to sit or stand erect would cause a recurrence of the vertigo, confusion of the head, and a sense which he described as if the sounds produced were far away, gradually receding as the vertigo became greater. Being a very intelligent observer, he was able to describe exactly the plane of gyration, which, if my recollection is correct, was in the direction of the posterior rather than of the perpendicular or horizontal semicircular canal. The treatment consisted of rest, the administration of *kali brom.;* and, after an absence from business more or less prolonged, all these symptoms disappeared without loss of hearing. This is a typical case of auditory or labyrinthine vertigo, and is undoubtedly functional in character.

Analogous to this, but giving results much more serious, is a primary idiopathic form of disease, very closely resembling cerebro-spinal meningitis, for which, in some cases, it has undoubtedly been mistaken. To this disease of the internal ear, Voltolini has given the term "otitis labyrinthica;" and Dr. Knapp the term, "otitis interna exudativa serosa." Some years since, I was called in consultation by Professor Doughty, to see a girl some ten or twelve years of age, who had suddenly lost her hearing while suffering from what he supposed, at the outset, to be an attack of cerebro-spinal meningitis. The child had all the symptoms of this dread disease, save the spinal tenderness to touch, and the convulsive action. On the third day she suddenly became deaf, so that, within a few hours, audition was perfectly obliterated. On examination, I found the condition as above stated, and expressed the opinion that it was a primary idiopathic disease of the labyrinth; the loss of function being due to serous exudation, and offering a very grave prognosis as regarded recovery of the function.

My friend, W. S. Searle, M.D., of Brooklyn, had given me, some time previously, a verbal account of a similar case, in which he believed *silicea* had been of great value in overcoming the serous exudation. The child was put upon *gelsemium* first, every hour or half-hour, and intercurrent doses of *silicea* every third hour.

Under these remedies, the fever, head-ache, occipital pain, and other constitutional symptoms, gradually disappeared. The hearing improved slowly but steadily, until the patient was able to hear as perfectly as ever. I think we are warranted in claiming this as an acute primary otitis interna exudativa; and the result obtained was due, I believe, to the immediate administration of remedies.

In two other similar cases, the loss of function was complete and permanent, the patients receiving no treatment directed especially to the preservation of the acoustic function. In one of these cases, the patient had suffered from suppurative disease of the middle ear on one side for many years, and in a summer evening, while playing croquet, had become very warm. Completing his part of the game, he retired to the veranda of a house near by, where he sat in his chair leaning against the house. He fell asleep, and was suddenly awakened by a confusion of noises which he was unable to distinguish. This was produced by the rest of the party shouting at the successful termination of their game. On attempting to rise, he found himself very dizzy, nauseated, and unable to control his steps sufficiently to walk to his room. He retired, passed a somewhat restless night, and, on waking in the morning, was alarmed at the excitement of his wife, who was attempting to convey to him her own alarm upon finding that he could not hear a single word which she had spoken to him. The history of the case would lead to the supposition that the hearing, up to that time, had been normal; as the patient had depended upon that ear for all ordinary conversation. Examination showed no evidence of middle-ear disease; and the hearing for watch, as well as the human voice, was absolutely *nil.*

In another instance (May 16, 1878), a lady, about thirty years of age, was taken suddenly with severe pain in the head, in the occipital region, nausea, giddiness, tinnitus aurium, and deafness of the right ear. She consulted her family physician, who did not recognize the gravity of the case, but thought her symptoms secondary to some gastric or hepatic derangement. After various means had been used, the nausea and giddiness gradually passed away; but the hearing did not improve. Six months later, on examination, I found the membrana tympani normal, the pharynx only slightly catarrhal, whereas the hearing on the right side was minus for the watch as well as for the tuning-fork.

On the left side the hearing was normal. The only cause which could be found was that of a general debilitated condition: otherwise, she was, and had been, in perfect health.

Not satisfied with my diagnosis and prognosis, she consulted eminent aurists in this city, who confirmed the diagnosis. This, like the other case, was undoubtedly a serous exudation in the labyrinth. It may seem assuming too much to claim, that, from the history, these were cases of serous disease; but I see no reason why, under the same method of treatment as already suggested, these cases might not have had at least a degree of functional activity preserved.

In cases of otitis interna suppurativa or hemorrhagica, the loss of hearing is sudden and permanent. That there may be a localized suppurative inflammation, as well as localized serous inflammation, it seems to me the reports of pathological examinations fully warrant. Otherwise, we cannot explain cases in which there is loss of portions of the musical scale. Theoretically, we infer, that, if a proportion of the lower tones is lost, the long turns of the cochlea are the portion involved: conversely, if the high tones are lost, the turns near the apex of the cochlea are involved. You can understand, that any exudations occurring in the labyrinth, in the tract of the acoustic nerve, or within the cranial cavity, which are absorbed without pseudo-organization, would, for the time being, cause a failure of the function, which might be gradually restored as the exudation was absorbed. If the exudation became purulent or hemorrhagic, you can understand that the disorganization resulting therefrom would be so serious that any restoration of the function of the part involved would be simply impossible. Such is the history of cases which we are led to suppose are either hemorrhagic or purulent. The same is true of inflammation of the internal ear, secondary to syphilis. This matter has been quite fully considered by various authors; and we find them divided into those who think that the preponderance of the lesion is found in the pharynx, and that otitis interna syphilitica is a very rare form of disease: hence, that which is considered as primary by some, is really secondary to disease of the middle ear and pharynx.

In February, 1878, I was called upon to advise a young man

who had suffered from catarrhal disease more or less for years, but never previously from any disturbance in the middle ear; nor was there, on examination, any evidence that the middle ear was involved more than in thousands of cases, — certainly not sufficiently to account for the sudden invasion, increasing to serious and permanent failure of the function. He had contracted syphilis two months previously, which had shown secondary symptoms of the skin and hair-bulbs, but no effects upon the mucous membrane until two weeks later than the ear-symptoms. Suddenly he was aware of slight dulness of hearing; subjective noises set in, the hearing became duller and duller; and, in his alarm, he was advised to consult me. As stated, I found no sufficient cause in the condition of the pharynx, Eustachian tube, middle ear: but examination with the tuning-fork showed perception by aerial conduction to be better than that through the cranial bones; and the test by closing alternately the right and left meatus, using the tuning-fork on the vertex, zygoma, or glabella, failed of the contrasts which we know to exist in cases in which the fault is one of conduction rather than of perception. *Iodide of potassium* and *mercurius* were of value in arresting the progress of the disease, but there was no improvement over the standard recorded at the first test.

In a number of instances, I have treated cases in which there was undoubtedly a complication of middle-ear disease and internal-ear disease; and I am ready to agree with those who would class such doubtful cases as secondary to middle-ear disease rather than as an exudation, and, later, as purulent processes secondary to the constitutional effects of the virus, independent of the middle-ear disease.

Otitis interna may be either primary or secondary, at the same time traumatic; resulting, if primary, directly from the shock or solution of the continuity of the labyrinth, independent of any middle-ear disease, or fracture of the temporal bone, or shock of the acoustic nerve itself. It may be secondary to disease of the middle ear set up by the same force which produced the shock, or secondary to traumatic causes which affect the auricle, temporal region, meatus, and tympanum directly, as from a blow immediately upon the parts. Fractures of the temporal bone, allowing the escape of the labyrinthine fluids, must necessarily result disastrously.

Traumatic causes generally affect the middle ear, giving rise to suppurative inflammation of the tympanum; frequently, by simple continuity of tissue, involve the internal ear, and destroy the function of the labyrinth. These all, on the grounds on which I have thus far considered the results, may be hypothetically accounted for by reason of the pathological processes; and there is a large group of cases in which the conditions have been classed as hyperæsthesia or torpor of the auditory function, which cannot be as satisfactorily explained.

To illustrate: in October, 1881, a clergyman consulted me for a distressing condition, which had existed for six years, and which he believed to be due to an attack of vertigo, so designated by himself, which occurred while he was specially overtaxed by mental strain. The attack was described as one of vertigo, with subjective noises, and a condition which he described "as if a cloud had settled down upon the function of the acoustic nerve, as a cloud might settle down upon the vision." Associated with this was a peculiar discomfort upon hearing the low, musical tones produced by any instrument, especially by the church-organ.

This particular symptom disappeared under the use of *chenopodium anthelminticum*, and he was very much improved by a mild current of galvanism. In this case, there was no absolute failure of the function, such as can be tested and recorded according to any of the usual standards; but there was a torpor and an unsatisfactory performance of the function, either of the terminal mechanism, or of the conduction of the acoustic nerve, or the central perceptive relation between sound and idea.

Like in kind with this, are those cases of hyperæsthesia in which the person is excessively sensitive to simple noises, regular vibrations in the air, or to some particular irregular vibrations, or to some regular tones, either high or low, or both, produced by musical instruments. In some cases, this is marked by a peculiarity which recognizes the color or quality, *klang farbe* (sound-color). These are analogous to those symptoms of the optic nerve in diseased conditions when the patient has illusions of color. That these are necessarily due to organic changes, I do not believe; that they do exist with organic changes, is undoubtedly true, as often the lesion is so great that the function is never restored. The trunk of the auditory nerve also is subject to changes in its structure, in the direction of atrophy, fatty degeneration, metamorphoses analogous to those occurring in the trunk of the optic nerve; also to malignant changes, which partially or wholly interfere with the function; and the same may be true of growths in the sheaths of the nerve, or in the periosteum lining the internal auditory canal.

You can readily understand how such processes would cause failure, either partial or complete, without interfering with the peripheral mechanism, being as absolute as when the cochlea and semicircular canals are the subjects of serous or purulent disease.

TWELFTH LECTURE.

DEAF-MUTISM: HELPS TO HEARING.

Gentlemen, — When the hearing has fallen below the normal standard, and internal remedies or instrumental treatment fail to increase it, we are compelled to resort to artificial methods for concentrating and conducting sound-waves. In speaking of the treatment of perforations, I mentioned the artificial membrana tympani and the cotton pellet, which are used as a substitute for the artificial membrane. These serve not only as a means of treatment, in many cases, promoting closure of the perforation, but, in cases where the perforation remains permanent, the artificial membrane, or the cotton pellet, acts to increase the hearing-power, by adjusting the relation of the ossicula, and thus conducts vibrations which otherwise would not be communicated to the nerve.

When the membrane has sloughed to a great extent, the relation of the malleus to the incus is so changed, that they are not in apposition, as in normal condition. This may be illustrated by imagining the cogs of wheels, in gearing, to be drawn apart, so that, instead of bearing fully in all their parts, the cogs touch, as it were, only on their ends. The action of the cotton pellet, or artificial drum-head, is similar to that of replacing the cogs in their proper position, so that the action of the gearing is easy and strong.

When the treatment of suppurative cases has progressed so far that the purulent secretion has ceased, the pledget of absorbent cotton may be rolled into a small ball, and placed so as to bear against the remaining portion of the manubrium of the malleus and the floor or walls of the canal. If properly adjusted, the hearing-distance will be markedly increased, unless adhesions have formed in such a way as to prevent the vibration of the ossicula. The same is true of the use of the artificial membrana tympani, but I have found that the cotton pellet is much better tolerated.

The degree of hearing which is secured by this little device is remarkable, — in some cases in the proportion of one to three, or higher, as one to ten.

In cases of chronic catarrhal inflammation, with thickening of the membrana tympani, adhesions of the same to the ossicula or labyrinthine wall, or possible pseudo-anchylosis of the stapes, some means must be used to concentrate and conduct the sound-waves : these consist of various forms of ear-trumpets. The conversation-tube consists of a flexible

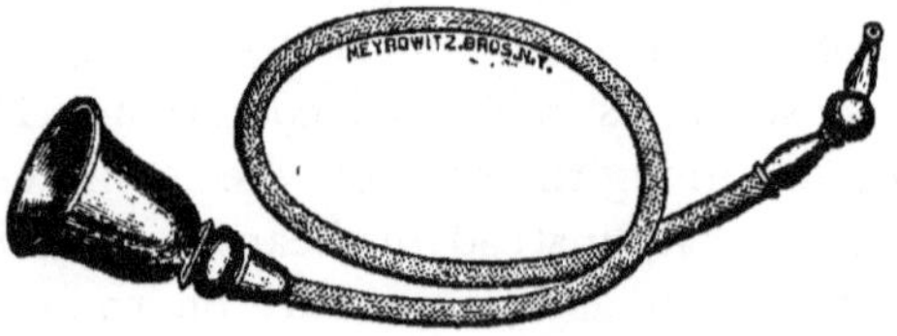

CONVERSATION-TUBE.

tube, finished at one end with a tip fitted to the external auditory canal, and at the other end with a bell-shaped mouth-piece of hard rubber. This instrument is very acceptable to some subjects, as it is not so cumbersome or noticeable as the large trumpet. The trumpets have been made in various forms. The best is the one constructed on the same principle as that underlying the construction of speaking-tubes, so that the air is thrown by vibration in right lines. All attempts at securing any increase of vibration by small telescopic devices, must necessarily prove failures. In cases of extreme deafness, the large so-called "dipper"-trumpet is the only serviceable instrument. This has been adapted by some instrument-makers so that it can stand on a base, with the mouth towards the speaker, the vibration being conveyed to the ear by a flexible tube. The best instrument of which I have any knowledge, constructed on these principles, is North's earphone, so called, which consists of a trumpet-shaped instrument, covered by meshes of fine wire, which the inventor claims to be of great value in modifying the vibrations. Whatever may be the theory upon which it is constructed, it is an admirable instrument.

The fact that a person can hear the orchestral music in the concert-room very much better by placing a cane upon the floor, and resting the head of it against the teeth, or by placing the brim of a Derby hat against the teeth, with the crown turned towards the stage, suggested, in a crude way, the principle which underlies the audiphone and the dentiphone. These instruments are alike in principle and construction, and consist of a sheet of vulcanite, fitted with a handle much like an ordinary fan. Placed against the edge of the teeth, with upward pressure on the handle, the instrument is bent in the form of an arc, with its base inward. It is fitted with silk cords passing from the upper edges to the handle, so that the tension may be regulated and preserved while in use. The statements of patients are not uniform as regards the use of these instruments, but people evidently require some training in order to become proficient in their use; and, as they are less unsightly than the ordinary trumpet, it will be well for a person requiring some aid to make a trial of their use. Dr. Knapp has recorded, in "The Archives of Otology," the comparative merits of the dipper-trumpet and the audiphone on very different patients, and does not say much in commendation of the audiphone.

EAR-TRUMPET.

The instruction of the deaf-mute is a matter upon which the advice of the general practitioner may be of such inestimable value to the sufferer, that I desire at this time to urge upon your attention the importance of methods of instruction, or, rather, what may be called improved methods of instruction, compared with those in vogue in past years. The child who is congenitally a deaf-mute, of course has no idea of language as expressed by the speaking-child; and the development of mental powers must necessarily depend very largely upon some substitute for this avenue of communication to the mind. In this matter, very great advances have been made. The old method of dactylology, or finger-signs, which has been taught for a long period, is being put in trial with another system, which has been in use in Europe for many years, more lately in this country. By this system, the finger-alphabet is prohibited, as well as all sign-language; and the children are taught, not only to understand others by watching the lips, but to use articulate speech themselves. This is acquired by patient, diligent training of the powers of imitation and observation. Every medical man is under obligations to inform himself, as far as possible, of the merits of these two systems, so that he may be able to give proper advice when his opinion is asked upon this subject.

Dr. W. B. Dalby gives a very full description of the so-called German method, in his lecture on "The Diseases of the Ear," as also in an article entitled, "The Education of the Deaf and Dumb by Means of Lip-Reading and Articulation." The so-called German method of the instruction of mutes may be briefly stated to consist in conveying to the child's mind the idea of the sounds which are associated with general positions of the lips and organs of speech.

Without going too fully into this subject, it is, perhaps, sufficient, in this connection, to say, that this training begins at seven years of age, the time when acquired deafness would result in deaf-mutism as truly as in the case of the congenital; and about eight years are occupied in the training, before the child can practise lip-reading, so called. By numerous objects necessary to attract the attention of the pupil, and by various signs, which the apt teacher can employ as best suited to his or her purpose, the child's intuition or perception of the relation of things is stimulated. When once the principle dawns upon the mind of the apt pupil, the mental enlightenment of one who has been hitherto a prisoner, finds, perhaps, its best description in Dickens's account of Dr. Howe's training of Laura Bridgman. The same patience and genius that led her soul out of bondage, will reap much larger harvests in the practical application of lip-reading to the necessities and enjoyment of every-day life.

Diseases of the Internal Ear.—Cases.

Ella W., aged 18, Elizabeth, N.J. *Otitis interna traumatica.* Admitted April 12, 1883. About two or three weeks since, fell fourteen steps, and struck upon a stone floor, thus injuring right side of head, and impairing sight and hearing; sent by Dr. Sprague to see Dr. Norton, about April 7; has had repeated hemorrhages from the right meatus and nose, five or six per day; examined also by Dr. Houghton; no bone conduction, faint aerial conduction.

April 12. Hearing gone, and numbness of right side, head and arm; two hemorrhages to-day, one from ear, one from nose.

April 13. One hemorrhage; progressive paralysis of motor nerves; retention of urine, drawn freely by catheter; abnormal position of the meatus urinarius.

April 14. Apparent complete paralysis of right arm; tongue thrust to right side; could not swallow; urine retained; one slight hemorrhage from nose in the afternoon; very bad night last night; delirious; temp., 99°, both sides, axillæ; seemed better in the forenoon, worse in the afternoon; no appetite; great thirst. *Ferrum phos.*

April 15, 1 P.M. Urine still retained; temp., 98°, left side; 99°, right side; excruciating pain in left side of head, coming in paroxysms, five or six in twenty-four hours, lasting from ten minutes to three-quarters of an hour; very bad night—little more rest after midnight; no appetite; thirst; has moved right arm to head; swallowed better; moved head to left side; previously unable to do so, on account of great pain and tenderness of left side of neck; falls into a drowse, with snoring, about one minute, and awakens with a start; projects tongue only half an inch, and it seems difficult to open mouth beyond half an inch. *Ferrum.*

April 16. Temp., right side, 99°; left side, 98.4°; one hemorrhage from nose at 1 P.M., not profuse; intellect bewildered, memory lost, etc.; fair night; worse this forenoon than at any other time; paroxysms of pain, convulsive startings still continue. *Bell.*

April 17. Several bad "spells" of pain; urination voluntary; menses come on; right arm moved while unconscious, thrown up over head, and back with slow, heavy motion; temp., right side, 99°; left side, 97.6°; no hemorrhages. *Hyos.*

April 18. Had hemorrhage from ear at 3 A.M., and from nose at 9 A.M.; crusted blood in L. meatus; the slightest touch causes pain on vertex; says she can tell when she is about to bleed, from sensation of fluid moving in head, gurgle or rattle; since hemorrhage, better of reflex symptoms of stomach and convulsive movements; temp., right side, 98°; left side, 97°.

April 19. Temp., right side, 97.3°; left side, 97°. 10 P.M. Temp., right side, 99°; left side, 98°. 10 A.M. Temp., right side, 99.3°; left side, 99°.

April 20. Doing well; dark hemorrhage from the ear last evening at 9 o'clock; rested well; no change in the paralysis motor or sensory. 11 P.M. Temp., right side, 99°; left side, 98°. 11 A.M. Temp., right side, 99.4°; left side, 99°.

April 21, 11 P.M. Temp., right side, 99°; left side, 98°.

April 22, A.M. Temp., right side, 99.4°; left side, 99°. 10 P.M. Right side, 99°; left side, 98°.

April 23, 11 A.M. Temp., right side, 99.4°; left side, 99°. 10 P.M. Right side, 100°; left side, 99.4°.

April 24, 11 A.M. Temp., right side, 99°; left side, 98°; much better; no hemorrhage; moves left arm and fingers; sensibility returning in arm, but not in face and neck; passes urine to-day all right. 11 P.M. Temp., right side, 98.4°; left side, 98°.

April 25, 10 A.M. Temp., right side, 98.3°; left side, 98°. 11 P.M. Right side, 99°; left side, 98°.

April 26, 11 A.M. Temp., right side, 98.4°; left side, 98°. 10 P.M. Temp., right side, 99°; left side, 98.4°.

April 27, 10 A.M. Temp., right side, 98.4°; left side, 98°. 10 P.M. Temp., right side, 98.4°; left side, 98°. 11 AM. Temp., right side, 99°; left side, 98.4°; sensation returned at 2 A.M., preceded by pain.

April 28. Much better; no special symptoms.

May 7. Had gained till 6th; head-ache, frontal, and running through side of head (R.); went out to walk at 10 A.M., then quiet till evening; went to church at 7.30; soon became dizzy; the lights seemed to go out; kept quiet until the close of services; on way to cross the street, was nauseated till near 10 P.M., when blood flowed from R. ear and R. nostril; then nausea ceased, and a beating sound, as of metal being struck, was heard; beats synchronous with pulse, R. *Me.*, crusted blood; R. *Mt.*, clear above, blood-scales below. *Ferrum.*

May 14. Was well last week, save nose-bleed, a slight amount daily; has hemorrhage from ear to-day at 10 A.M.; feels chilly, and seems nervous; pulse normal; seems well in other respects.

May 16. Hemorrhage last night, again since 12 M.; meatus tender to the pressure of speculum; blood coagula in the lower inner third of meatus. *Hepar.*

May 18. Four hemorrhages since last visit, three from ear, one from nose, dark and thick; nervous; sighing respiration; restless nights. *Ign., ferrum phos.*

May 21. Had five "bleeding-spells" on the 19th, one last evening about 8 o'clock, and one this forenoon; "sees stars" with right eye, on closing left; less nervous; sleeps better; less restless. 9 P.M. Had a bad nervous attack this morning, teeth clinched, extremities cool, pulse slow and small; had been having several hemorrhages, through the day, of dark, thin blood. *Ergot* and *ignat.*

May 22. Better this morning; no hemorrhage; head-ache

on right side of face, and across forehead; right side of face flushed. *Bell.* Hemorrhage from nostrils and meatus at 2.30. *Ipec.*

May 23. Had hemorrhage this morning; has a dark clot in meatus, and walls look as if abraded; *Mt.* more tender. *Ham., hyos.*

May 25. Bled five times yesterday, quite profusely, from ear and nose; blood still dark. *Ham., hyos.*

May 31. Has the hemorrhage when she goes by herself into the water-closet; has had only one at any other time. *Ferrum, hyos.*

June 1. No hemorrhage since the 28th. *Ferrum, hyos.* Menses appeared on the 28th, one day, dark and very scanty. *Ferrum, hyos.*[1]

Case No. 21. — Alice Chamberlain, aged twenty-one years. *Otitis interna.* Dec. 20, 1882. Three years ago her vision began to fail. Two years later, the hearing became impaired. She has been subject to frequent and prolonged attacks of head-ache. The failure of hearing was first noticed as asso-

[1] June 25. The patient was discharged at the above date, and went to her former employer. To-day she called at my office. As the menses are about to appear, the head is more confused, and blood flows from the meatus, but not from the nose; the floor of the canal is red, sensitive, and covered with light-colored bloody fluid; no point of rupture in canal or *Mt.* can be seen; the general condition is better than one month ago.

This case is interesting, because of so good a degree of restoration. Prognosis was guarded, and the diagnosis was not clear till the removal of tenderness of cervical muscles of left side was followed by restored motor and sensory function of the right side. Brown-Sequard mentions some cases of central lesion with paralysis of same side. Dr. St. Clair Smith called my attention to cases in Archives of Scientific and Practical Medicines, February, 1878. It seems clear that the blow upon the temporal bone, which caused the lesion of the internal ear, also caused temporary failure of the function of both anterior and posterior cords of cervical nerves; this giving a condition which caused much solicitude and debate As regards the future history of the case, I anticipate a gradual repair of the lesion of the temporal, and cessation of the hemorrhage.

May 25, 1885. Ella W. returned to the New-York Ophthalmic Hospital a few days ago, and the following additional facts were learned: Soon after returning to service, the hemorrhages returned. She was admitted to some hospital in Elizabeth, at the request of the priest. There she had numerous hemorrhages; anæsthesia of right side; the skin became dark, and remained so for weeks. In October she was able to sail for Ireland. During the voyage she was very sick, and for two weeks after landing: from that time she made rapid progress. In November the flow of blood ceased entirely, and she seems in perfect health. There is complete absence of aerial and bone conduction.

ciated with a constant noise, as the rolling of barrels, the sounds of moving wagons, near her. The same morbid process, which progressed two years in the optic tract, had now undoubtedly commenced its ravages in the auditory trunk. The persistent head-aches to which she was subject must be considered a grave symptom when associated with either the optic or auditory nerve disease. It is only by analyzing any given case, including its history, that you will be able to determine whether the head-ache is due to centric causes, or symptomatic of some disease of the organism, possibly of some remote part. This case was clearly an idiopathic disease. The remedy which arrested the neuritis was *spigelia*, prescribed by Professor George S. Norton, M.D., from whose clinic she came to me. The tinnitus aurium has increased during the past year, and is associated with a sensation of vertigo, especially on raising the head after waking in the morning, the bed seeming to revolve in a horizontal plane.

March, 1883. This patient has improved steadily under the use of spigelia: the optic neuritis has been arrested, the vertigo has passed, the tinnitus aurium has been reduced to a minimum, and every thing indicates the arrest of the morbid process which threatened to abolish both functions.

Case No. 18.—Bella Bronson, two and a half years of age. *Otitis interna secondaria.* Dec. 20, 1882. When this child was ten weeks old, she was taken with fever of some type, and, so far as the history of the case can be determined, recovered without apparent trace of brain-trouble. She came to the clinic yesterday, possessing the same degree of natural brightness and joyous activity which you see her present to-day; but I am sorry to say that she is afflicted with disease of the internal ear. In this case, as you will actually find, the diagnosis is reached by exclusion. Examination reveals that the external and middle ear are normal: therefore the defective hearing must be due to disease of the internal ear. She has enjoyed perfect health. The tests that have been applied show that she gets some idea of sound; but whether by vibrations communicated to her by aerial waves, or whether by the shock conveyed to the body through contact with the floor, it is impossible to determine in so young a child: I believe, though, that the child is a deaf-mute, as the result of fever when she was ten weeks of age.

I now recall the case of the girl, ten years of age, who was taken with what was believed to be cerebro-spinal meningitis. On the fourth day of her illness she was attacked with sudden loss of hearing. The case was under the care of Dr. Doughty, who called me in consultation. After examination, I determined that the disease was not one of the middle ear, nor were there evidences of spinal or cerebral lesion, as in cerebro-spinal meningitis. She was placed under the influence of *gelsemium*, with intercurrent doses of *silicea*. The gelsemium was given on the marked indications of the pulse, and the head-symptoms as given by the child; and the silicea on the presumption that the pathological conditions were as noted by Dr. Searle of Brooklyn, he having relieved a similar case of otitis interna serosa with that remedy. Immediate improvement was manifested: in twenty-four hours she had improved; in ten days the hearing-power had been restored to a very great degree; in six weeks, there was a normal condition of the patient, and perfect hearing. I refer to this case again, in order to impress upon you the importance of detecting the primary symptoms, so as to guide the differential diagnosis between otitis interna and cerebro-spinal meningitis, which shall avert the progress of the disease. The same testimony of the importance of the early recognition of this condition will be found in the record of cases reported in this country and in Europe.

Professor H. Knapp, M.D., remarks, that, "when the serous exudation passes beyond the acute stage into that of pseudo-organization, the case is hopeless."

I am satisfied, that, in this case, the serous exudation was controlled, and destruction of the terminal nerve-filaments was averted, by the prompt administration of the remedies.

Summary of Remedies.

Chenopodium anthelminticum. — "Deafness to the sound of the voice, but exquisite sensitiveness to the sounds of passing vehicles; he remarked, as each vehicle rolled by, that it sounded like the roaring of immense cannons right into his ear; also annoying buzzing in ears. During all this time, his deafness, as described, was progressive, and became so pronounced as to make it impossible to talk to him. Still, there was the same kind of sensitiveness to other sounds. For example, when the tea-bell rang, though he was in the third story, three flights from where the sound came, he, without notice from members of his family, to their utter astonishment, got up and walked, as deliberately as ever, into the dining-room."

Cinchona, and especially its alkaline proximate principle, quinia, has been long recognized as having toxic effect on the ear. The symptoms produced by the proving, point to impressions upon the labyrinth, such as produced by sudden abstraction of blood; viz., vertigo, from loss of blood; giddiness, from anæmia; fainting; ringing in ears. In some subjects, a fine ringing in the ears is caused, associated with a general nervous erethism. As a remedy, *cinchona* proves curative in these two opposite conditions, and acts upon both cochlea and semicircular canals.

Roosa[1] gives decided testimony as to the effect of quinine. "I believe that the tinnitus aurium, and impairment of hearing, following the use of quinine, depend upon congestion of the ultimate fibres of the auditory nerve in the cochlea; and that the redness of the drum-heads is merely an index of the former condition." He quotes Kirchner to this effect: "*Quinine causes inflammatory processes and permanent pathological changes in the ear. He believes that the cause for these conditions is to be found, not only in a hyperæmia of short duration, but also in paralysis of the vessels, with congestion and exudation.*" Also, Dr. J. Orne Green, as follows:

[1] Treatise on Diseases of the Ear, p. 619. New York, 1885.

"*From our present knowledge, both clinical and experimental, we are justified in asserting, that the action of quinine upon the ears is to produce congestion of the labyrinth and tympanum, and sometimes distinct inflammation, with permanent tissue-changes.*"

Thus the modern scholars give reasons on a pathological basis, for the truth of observations made by Samuel Hahnemann just one century ago. We shall still need to follow his lead, to solve the action of cinchona.

Ferrum phosphoricum. — For the same general symptoms as in middle-ear disease; i.e., the first stages of inflammation. (See Otitis Media.)

Gelsemium. — In otitis interna serosa. (See symptoms under Otitis Media.)

Hydrobromic acid. — There is no proving of this remedy: hence our use of it has been guided by the cases reported by Drs. Woakes, Turnbull, and others. Like the bromides, it is a strong sedative, in doses of from five to thirty drops. A convenient method is to put thirty drops in three tablespoonfuls of cool water; add sugar to suit the taste, and take a tablespoonful each hour. In pulsating tinnitus, with great nervous irritability, it has done me service. In one case, it produced excessive irritability, and the patient abandoned it.

Pilocarpin muriate. — Politzer reports effects from this salt when injected hypodermically in the mastoid. We had the low potencies prepared, and obtained some decided results. One child, who was a deaf-mute from some acute disease, has gained, by slow stages, through three months' use of the remedy. At first, there was produced an extreme sensitiveness to very loud sounds; later, a perception of lower sounds, unrecognized previously; till now, most sounds are perceived, if quite loud. With this, language is returning; and the child now understands nearly every command addressed to her, as is shown by the articles brought and carried as directed.

An adult patient showed similar effects. Last May he

took a cold bath while in profuse perspiration: inside of twelve hours he lost his hearing suddenly, with nausea, vomiting, dizziness on attempting to rise from the bed. After confinement to the house for four or five weeks, he began gradually to walk out of doors, but staggered as if intoxicated.

In October, 1884, when examined by me, his hearing for the watch was c-20 R., -20 L.; tuning-fork before ear, not on bones or teeth, feels the jar; heard bells, etc., but could not distinguish voice unless very near the auricle; heard metronome 4-20 R., 2-20 L.; he still staggered at times; tinnitus like roar of a mill, has a very depressing effect; thinks and talks of nothing else.

Muriate of pilocarpin increased the hearing, but also increased the tinnitus: no argument would induce the patient to endure the increased tinnitus for the sake of prospective power. The remedies for lesions of the internal ear are so few in number, that any new one will be of special interest, and worthy of thorough test.

Salicylic acid and its salts, *salicylate of potash*, and *salicylate of soda*, are all known to produce lesions of the labyrinth: for want of a proving, we depend on reports of effects of massive doses; these are confirmed by reports of clinical observations, as we find that auditory vertigo is relieved by these remedies; in one case, salicylate of soda relieved a child, who had a clear history of otitis interna exudativa serosa. (See Quinine.)

Cured Symptoms.

Mercurius. — Coppery odor of pus from the meatus and cavity of the tympanum. Chronic inflammation and suppuration of the drum-head.

Mercurius dulcis. — Closure of the Eustachian tube, with the deep-red color of the mucous membrane of the posterior wall of the pharynx characteristic of granular pharyngitis.

Mercurius dulcis. — A deep-toned roaring, in cases of chronic catarrh of the middle ear, with closure of the Eustachian tube. This remedy is particularly valuable in progressive loss of hearing in old people, when it is based on chronic naso-pharyngeal catarrh with granular pharyngitis.

Hepar sulph. calc. — Extreme sensitiveness of the meatus and drum-head in the otitis media suppurativa chronica, with perforation of the drum-head.

Iodine. — Roaring as of a mill, in case of chronic catarrhal inflammation of the middle ear, such as is sometimes classed as proliferous or sclerosis.

Kali mur. — Closure of the Eustachian tube in chronic suppurative inflammation of the middle ear.

APPENDIX.

LIST OF ABBREVIATIONS.

Ab. c., Abies canadensis.
Ab. n., Abies nigra.
Abs., Absinthium.
Acal., Acalypha indica.
Ac. ac., Acetic acid.
Aconin., Aconitine.
Acon. a., Aconitum anthora.
Acon. c., Aconitum cammarum.
Acon. f., Aconitum ferox.
Acon. l., Aconitum lycoctonum.
Acon., Aconitum napellus.
Acon. s., Aconitum septentrionale.
Act., Actæa spicata.
Adel., Adelheidsquelle.
Adox., Adoxa.
Æs. g., Æsculus glabra.
Æs. h., Æsculus hippocastanum.
Æth., Æthusa.
Aga. campn., Agaricus campanulatus.
Aga. camps., Agaricus campestris.
Aga. cit., Agaricus citrinus.
Aga. em., Agaricus emeticus.
Aga. m., Agaricus muscarius.
Aga. pa., Agaricus pantherinus.
Aga. ph., Agaricus phalloides.
Aga. pr., Agaricus procerus.
Aga. se., Agaricus semiglobatus.
Aga. st., Agaricus stercorarius.
Agk., Agkistrodon contortrix.
Agn., Agnus castus.
Agro., Agrostemma githago.
Ail., Ailanthus.
Alco., Alcohol.
Ald., Aldehyde.
Alet., Aletris farinosa.
All. c., Allium cepa.
All. s., Allium sativum.
Aloe, Aloe.
Alst., Alstonia scholaris.
Alum., Aluminæ.
Alumn., Alumen (alum).
Ambra, Ambra.
Ambro, Ambro artemisiæfolia.
Ammc., Ammoniacum (gum amm.).
Am. ac., Ammonium aceticum.
Am. be., Ammonium benzoicum.
Am. br., Ammonium bromidum.
Am. car., Ammonium carbonicum.
Am. cau., Ammonium causticum (ammonia).
Am. i., Ammonium iodatum.
Am. m., Ammonium muriaticum.
Am. n., Ammonium nitricum.
Am. p., Ammonium phosphoricum.
Ampe., Ampelopsis.
Amph., Amphisbœna.
Amyg., Amygdalæ amaræ aqua.
Aml. n., Amyl nitrate.
Aml. ch., Amylamine chlorohydrate.
Anac., Anacardium.
Anag., Anagallis.
Anan., Anantherum.
Ange., Angelica atropurpurea.
Angu., Angustura.
Anil., Anilinum.
Anis., Anisum stellatum.

Anth. n., Anthemis nobilis.
Anthr., Anthrakokali.
Ant. a., Antimonium arsenitum.
Ant. m., Antimonium mur. (chlor.).
Ant. cr., Antimonium crudum.
Ant. ox., Antimonium oxidum.
Ant. s., Antimonium sulf. auratum.
Ant. t., Antimonium et. potass. tart.
Apis, Apis.
Ap. g., Apium graveolens.
Aph., Aphis chenopodii glauci.
Apoc. a., Apocynum andros.
Apoc. c., Apocynum cannab.
Apom., Apomorphine.
Aq. m., Aqua marina.
Aq. p., Aqua petra.
Aral., Aralia racemosa.
Aran., Aranea.
Aran. d., Aranea diadema.
Aran. s., Aranea scinencia.
Argem., Argemone.
Arg. c., Argentum cyanidum.
Arg., Argentum metallicum.
Arg. mu., Argentum muriaticum.
Arg. n., Argentum nitricum.
Arist. c., Aristolochia columb.
Arist. m., Aristolochia (milhomens).
Arist. s., Aristolochia serpentaria.
Arn., Arnica.
Ars., Arsenicum album.
Ars. h., Arsenicum hydrogenisatum.
Ars. i., Arsenicum iodatum.
Ars. m., Arsenicum metallicum.
Ars. s. f., Arsenicum sulf. flavum.
Ars. s. r., Arsenicum sulf. rubrum.
Art. ab., Artemisia abrotanum.
Art. v., Artemisia vulgaris.
Arum d., Arum dracontium.
Arum i., Arum italicum.
Arum m., Arum maculatum.
Arum t., Arum tryphyllum.
Arun. Arundo mauritanica.
Asaf., Asafœtida.
Asar., Asarum.
Asc. c., Asclepias cornuti (Syriaca).
Asc. t., Asclepias tuberosa.
Asim., Asimina triloba.
Asp., Asparagus.
Ast., Asterias rubens.
Astr., Astragalus Menziesii.
Ath., Athamanta.
Atro., Atropinum.
Aur., Aurum.
Aur. ful., Aurum fulminans.
Aur. m., Aurum muriaticum.
Aur. m. n., Aurum muriaticum natro.
Aur. s., Aurum sulfuratum.

Bad., Badiaga.
Bals., Balsamum Peruvianum.
Bap., Baptisia.
Bart., Bartfelder (acid spring).
Bar. ac., Baryta acetica.
Bar. c., Baryta carbonica.
Bar. m., Baryta muriatica.
Bell., Belladonna.
Bell. p., Bellis perennis.
Benz., Benzinum.
Benz. n., Benzinum nitricum.
Benz. ac., Benzoic acid.
Benzoin, Benzoin.
Berbn., Berberinum.
Berb., Berberis.
Bism., Bismuthum oxidum.
Bla., Blatta Americana.
Bol. l., Boletus laricis.
Bol. l., Boletus luridus.
Bol. s., Boletus satanus.
Bom. c., Bombyx chrysorrhœa.
Bom. p., Bombyx processionea.
Bon., Bondonneau.
Bor., Borax.
Bor. ac., Boracicum acidum.
Bot., Bothrops lanceolatus.
Bou., Bounafa.
Bov., Bovista.
Brach., Brachyglottis.
Brass., Brassica napus.
Bro., Bromium.
Bru., Brucea antidysenterica.
Brucn., Brucinum.
Bry., Bryonia.
Buf., Bufo.
Buf. s., Bufo sahytiensis.
Bux., Buxus.

Cact., Cactus.
Cad., Cadmium sulfuratum.
Cad. b., Cadmium bromatum.
Cai., Cainca.
Caj., Cajuputum.
Cala., Caladium.
Calc. a., Calcarea acetica.
Calc. c., Calcarea carbonica.
Calc. cau., Calcarea caustica.
Calc. cl., Calcarea chlorata.
Calc. f., Calcarea fluorata.
Calc. i., Calcarea iodata.
Calc. m., Calcarea muriatica.
Calc. p., Calcarea phosphorica.
Calc. s., Calcarea sulfurica.
Cale., Calendula.
Calo., Calotropis.
Calth., Caltha.
Camph., Camphora.
Canc. f., Cancer fluviatilis.
Canch., Canchalagua.
Canna, Canna.
Cannab. i., Cannabis indica.
Cannab. s., Cannabis sativa.
Canth., Cantharis.
Cap., Capsicum.
Carb. a., Carbo animalis.
Carb. v., Carbo vegetabilis.
Carb. ac., Carbolic acid.
Carbn., Carboneum.
Carbn. cl., Carboneum chloratum.
Carbn. h., Carboneum hydrogenisatum.
Carbn. o., Carboneum oxygenisatum.
Carbn. s., Carboneum sulfuratum.
Card. b., Carduus benedictus.
Carl., Carlsbad.
Caru., Carum.
Cary., Carya alba.
Casc., Cascarilla.
Cass., Cassada.
Cast. v., Castanea vesca.
Cast. eq., Castor equi.
Castor., Castoreum.
Cata., Catalpa.
Caul., Caulophyllum.
Caus., Causticum.
Ced., Cedron.
Celt., Celtis.
Cent., Centaurea tagana.
Ceph., Cephalanthus.
Cer. b., Cereus bonplandii.
Cer. s., Cereus serpentinus.
Cerv., Cervus.
Cham., Chamomilla.
Chel., Chelidonium majus.
Chen. a., Chenopodium anthel.
Chen. v., Chenopodium vulvaria.
Chim., Chimaphila.
Chin., China.
Chin. a., Chininum arsen.
Chin. m., Chininum muriat.
Chin. s., Chininum sulfuric.
Chio., Chionanthus.
Chlol., Chloralum.
Chlf., Chloroformum.
Chlo., Chlorum.
Chr. ac., Chromium acidum.
Chr. ox., Chromium oxidatum.
Chry. ac., Chrysophanic acid.
Cic., Cicuta maculata.
Cic. v., Cicuta virosa.
Cich., Cichorium.
Cimx., Cimex.
Cimic., Cimicifuga.
Cina, Cina.
Cinch., Cinchoninum sulfuratum.
Cinnb., Cinnabaris.
Cinnm., Cinnamomum.
Cist., Cistus.
Cit. ac., Citric acid.
Cit. l., Citrus limonum.
Cit. v., Citrus vulgaris.
Cle., Clematis.
Cob., Cobaltum.
Coca, Coca.
Cocci., Coccionella.
Cocc., Cocculus.
Coc. c., Coccus cacti.
Coch., Cochlearia.
Cod., Codeinum.
Coff., Coffea cruda.
Coff. t., Coffea tosta.
Coffn., Coffeinum.
Colch., Colchicinum.
Colchm., Colchicum.
Colocn., Colocynthinum.

Coloc., Colocynthis.
Coll., Collinsonia.
Com., Comocladia.
Conch., Conchiolinum.
Con., Conium.
Conin., Coniinum.
Conin. br., Coniinum bromatum.
Conv. a., Convolvulus arvensis.
Conv. d., Convolvulus duartinus.
Cop., Copaiva.
Coral., Corallium rubrum.
Cori. m., Coriaria myrtifolia.
Cori. r., Coriaria ruscifolia.
Corn. c., Cornus circinata.
Corn. f., Cornus florida.
Corn. s., Cornus sericea.
Cot., Cotyledon.
Croc., Crocus.
Crot. h., Crotalus horridus.
Crot. c., Crotalus cascavella.
Crot. t., Croton tiglium.
Cry., Cryptopinum.
Cub., Cubeba.
Cun., Cundurango.
Cupress, Cupressus Australis.
Cup., Cuprum.
Cup. ac., Cuprum aceticum.
Cup. am., Cuprum ammonio-sulfu.
Cup. ars., Cuprum arsenicosum.
Cup. m., Cuprum muriaticum.
Cup. n., Cuprum nitricum.
Cup. s., Cuprum sulfuricum.
Cur., Curare.
Cyc., Cyclamen.

Daph., Daphne Indica.
Dat. a., Datura arborea.
Dat. f., Datura ferox.
Dat m., Datura metal.
Dat. s., Datura sanguinea.
Delph., Delphinus.
Delphn., Delphininum.
Dem., Dematium petræum.
Der., Derris pinnata.
Dic., Dictamus.
Dig., Digitalis.
Dign., Digitalinum.
Digtx., Digitoxinum.
Dios., Dioscorea.
Dir., Dirca palustris.
Dol., Dolichos pruriens.
Dor., Doryphora.
Dro., Drosera.
Dub., Duboisia.
Dub. h., Duboisia hopwoodi.
Dul., Dulcamara.

Eaux, Eaux Bonnes.
Ech., Echites.
Elap., Elaps.
Elat., Elaterium.
Elæ., Elæis guineensis.
Eme., Emetinum.
Epi., Epilobium palustre.
Equ., Equisetum.
Ere., Erechthites.
Erig., Erigeron.
Ery. a., Eryngium aquaticum.
Ery. m., Eryngium maritinum.
Eryth., Erythrophlæum.
Erio., Eriodictyon.
Ese., Eserinum.
Eth., Ether.
Eth. n., Ethyl nitrate.
Euc., Eucalyptus.
Eug., Eugenia jambos.
Euo., Euonymus Europæus.
Eup. per., Eupatorium perfoliatum.
Eup. pur., Eupatorium purpureum.
Euph. a., Euphorbia amygdaloides.
Euph. co., Euphorbia corollata.
Euph. cy., Euphorbia cyparissias.
Euph. h., Euphorbia hypericifolia.
Euph. i., Euphorbia ipecacuan.
Euph. l., Euphorbia lathyris.
Euph. p., Euphorbia peplus.
Euphm., Euphorbium.
Euphr., Euphrasia.
Eupi., Eupion.

Fago., Fagopyrum.
Fagu., Fagus.
Fel, Fel tauri.
Fer., Ferrum.
Fer. i., Ferrum iodatum.
Fer. ma., Ferrum magneticum.

Fer. mu., Ferrum muriaticum.
Fer. p., Ferrum phosphoricum.
Fer. s., Ferrum sulfuricum.
Fer. t., Ferrum tartaricum.
Feru., Ferula glauca.
Fil., Filix mas.
Fl. ac., Fluoricum acidum.
Fœn., Fœniculum.
For., Formica.
Frag., Fragaria.
Fran., Franzensbad.
Frax., Fraxinus.
Fuc., Fucus vesiculosus.
Fuch., Fuchsin.

Gad., Gadus morrhua.
Gal. ac., Gallicum acidum.
Gam., Gambogia.
Gas., Gastein.
Gau., Gaultheria.
Gel., Gelsemium.
Gen., Genista.
Geo., Geoffroya.
Gent. c., Gentiana cruciata.
Gent. l., Gentiana lutea.
Ger., Geranium maculatum.
Get., Gettysburg.
Geu., Geum rivale.
Gin., Ginseng.
Glo., Glonoin.
Gna., Gnaphalium.
Gos., Gossypium.
Gran., Granatum.
Grap., Graphites.
Grat., Gratiola.
Grin., Grindella.
Guan., Guano.
Guara., Guarano.
Guare., Guarea.
Guai., Guaiacum.
Gymne., Gymnema.
Gymno., Gymnocladus.

Hæm., Hæmatoxylon.
Hal., Hall.
Ham., Hamamelis.
Hede., Hedeoma.
Hedy., Hedysarum ildefon.
Helia., Helianthus.
Helio, Heliotropinum.
Hell., Helleborus niger.
Hell. f., Helleborus fœtidus.
Hell. o., Helleborus Orientalis.
Hell. v., Helleborus viridis.
Helo., Helonias.
Hep., Hepar sul. calc.
Hepat., Hepatica.
Her., Heracleum.
Hip., Hippomanes.
Hom., Homeria.
Hur., Hura Brasiliensis.
Hydrs., Hydrastis.
Hydrc., Hydrocotyle.
Hyc. ac., Hydrocyanic acid.
Hydrphb., Hydrophobinum.
Hydrphl., Hydrophyllum.
Hyos., Hyoscyamus.
Hyosn., Hyoscyaminum.
Hype., Hypericum.
Hypo., Hypophyllum.

Ib., Iberis.
Ig., Ignatia.
Ill., Illicium.
Imp., Imperatoria.
Ind., Indium metallicum.
Indg., Indigo.
Inu., Inula.
Iod., Iodum.
Iodof., Iodoformum.
Ip., Ipecacuanha.
Ir. fl., Iris florentina.
Ir. fœ., Iris fœtidissima.
Ir. v., Iris versicolor.
Itu, Itu.

Jab., Jaborandi.
Jac., Jacaranda.
Jal., Jalapa.
Jas., Jasminum.
Jat., Jatropha.
Jat. u., Jatropha urens.
Jug. c., Juglans cinerea.
Jug. r., Juglans regia.
Junc., Juncus.
Juni., Juniperus Virginiana.

K. ac., Kali aceticum.
K. ar., Kali arsenicosum.
K. bi., Kali bichromicum.
K. br., Kali bromatum.
K. ca., Kali carbonicum.
K. chr., Kali chromicum.
K. clc., Kali chloricum.
K. cls., Kali chlorosum.
K. cy., Kali cyanatum.
K. fcy., Kali ferrocyanatum.
K. iod, Kali iodatum.
K. ma., Kali hypermanganicum.
K. n., Kali nitricum.
K. ox., Kali oxalicum.
K. pi., Kali picricum.
K. slft., Kali sulfuratum.
K. slfc., Kali sulfuricum.
K. tar., Kali tartaricum.
K. tel., Kali telluricum.
Kalm., Kalmia.
Kar., Karaka.
Kat., Katipo.
Kcrsln., Kerosolene.
Kersn., Kerosenum.
Kin., Kino (Australiensc).
Kis., Kissengen.
Kou., Kousso.
Kre., Kreosotum.

Lab., Laburnum.
Lace., Lacerta.
Lach., Lachesis.
Lachn., Lachnanthes.
Lac. ac., Lactic acid.
Lact., Lactuca.
Lam., Lamium.
Lap., Lapathum.
Lapp., Lappa.
Lath., Lathyrus.
Lau., Laurocerasus.
Led., Ledum.
Lepi., Lepidium.
Lept., Leptandra.
Lil. s., Lilium superbum.
Lil. t., Lilium tigrinum.
Lim., Limulus.
Lina., Linaria.
Linu., Linum.
Lip., Lippespringe.
Lith. c., Lithium.
Lith. m., Lithium muriaticum.
Lob. c., Lobia cardinalis.
Lob. i., Lobia inflata.
Lob. s., Lobia syphilitica.
Lobln., Lobelinum.
Lol., Lolium.
Lon., Lonicera.
Luf., Luffa.
Lup., Lupulus.
Lyc., Lycopodium.
Lycper., Lycopersicum.
Lycps., Lycopus.

Mac., Macrotinum.
Mag. c., Magnesia carbonica.
Mag. m., Magnesia muriatica.
Mag. s., Magnesia sulfurica.
Magn., Magnolia glauca.
Manc., Mancinella.
Mand., Mandragora.
Mang., Manganum.
Mang. m., Maganum muriaticum.
Mang. ox., Maganum oxidatum nati.
Mang. s., Maganum sulfuricum.
Mar., Marum verum.
Mat., Maté.
Mec., Meconium.
Med., Medusa.
Mela., Melastoma.
Meli., Melilotus.
Meni., Menispermum.
Ment. pi., Mentha piperita.
Ment. pu., Mentha pulegium.
Meny., Menyanthes.
Mep., Mephitis.
Merc., Mercurius.
Merc. ac., Mercurius aceticus.
Merc. b., Mercurius bromatus.
Merc. c., Mercurius corrosivus.
Merc. cy., Mercurius cyanatus.
Merc. d., Mercurius dulcis.
Merc. i. f., Mercurius iod. flavus.
Merc. i. r., Mercurius iod. ruber.
Merc. m., Mercurius methylenus.
Merc. n., Mercurius nitrosus.
Merc. p. a., Mercurius precipitatus albus.

Merc. p. r., Mercurius precipitatus ruber.
Merc. sol., Mercurius solubis.
Merc. sulcy., Mercurius sulfocyanatus.
Merc. sulf., Mercurius sulfuricus.
Merl., Mercurialis.
Methln., Methylenum bichloratum.
Meth. e. e., Methyl-ethyl ether.
Mez., Mezereum.
Mil., Millefolum.
Mim., Mimosa.
Mit., Mitchella.
Mom., Momordica.
Mon., Monotropa.
Mor., Morphinum.
Mos., Moschus.
Murx., Murex.
Mur. ac., Muriaticum acidum.
Muru., Mururc.
Musa., Musa.
Musn., Muscarin.
Myg., Mygale.
Myric., Myrica.
Myris., Myristica.

Na. ar., Natrum arsenicum.
Na. br., Natrum bromatum.
Na. c., Natrum carbonicum.
Na. hy., Natrum hypochlorosum.
Na. la., Natrum lacticum.
Na. m., Natrum muriaticum.
Na. n., Natrum nitricum.
Na. p., Natrum phosphoricum.
Na. sa., Natrum salicylicum.
Na. sl., Natrum sulfovinicum.
Na. slfc., Natrum sulfuricum.
Nab., Nabalus.
Naj., Naja.
Nap., Naphtha.
Narcn., Narceinum.
Narcot., Narcotinum.
Narcs. po., Narcissus.
Nars. ps., Narcissus pseudo-narcissus.
Narz., Narzan.
Nicc., Niccolum.
Nico., Nicotinum.
Nit. d. s., Nitri dulcis spiritus.
Nit. ac., Nitricum acidum.
Nit. m. ac., Nitro muriatic acid.
Nit. ox., Nitrogenium oxygenatum.
Nitr., Nitrum.
Nup., Nuphar luteum.
Nx. m., Nux moschata.
Nx. v., Nux vomica.
Nym., Nymphæa odorata.

Oci., Ocinum.
Oena., Oenanthe.
Oeno., Oenothera.
Olnd., Oleander.
Ol. an., Oleum animale.
Ol. j., Oleum jecoris aselli.
Oni., Oniscus.
Op., Opium.
Opl., Oplia.
Opu., Opuntia.
Ori., Origanum.
Osm., Osmium.
Ost., Ostrya.
Ox. ac., Oxalicum acidum.
Oxeo., Oxeodaphne.
Ozo., Ozonum.

Paeo., Paeonia.
Pal., Palladium.
Pan., Panacea.
Pap., Papaverinum.
Par., Paris quadrifolia.
Pas., Pastinaca.
Pau. p., Paullinia pinnata.
Pau. s., Paullinia sorbilis.
Pb., Plumbum.
Pb. ch., Plumbum chromicum.
Ped, Pediculus.
Pen., Penthorum.
Per., Persica.
Peti., Petiveria.
Petrol., Petroleum.
Petros., Petroselinum.
Phal., Phallus.
Phas., Phaseolus.
Phe., Phellandrium.
Pho., Phosphorus.
Pho. ac., Phosphoricum acidum.
Pho. h., Phosphoricum hydrogenatus.
Physa., Physalia.
Physo., Physostigma.

Phyt., Phytolacca.
Pic. ac., Picricum acidum.
Picro., Picrotoxin.
Pil., Pilocarpinum.
Pime., Pimenta.
Pimp., Pimpinella.
Pin. c., Pinus cupressus.
Pin. l., Pinus lambertiana.
Pin. s., Pinus sylvestris.
Pip. m., Piper methysticum.
Pip. n., Piper nigrum.
Pis., Piscidia.
Pix., Pix liquida.
Plan., Plantago.
Plat., Platinum.
Plat. m., Platinum muriaticum.
Ple., Plectranthus.
Plumbg., Plumbago littoralis.
Pod., Podophyllum.
Polyg., Polygonum.
Polyp. o., Polyporus officinalis.
Polyp. p., Polyporus pinicola.
Pop., Populus.
Pot., Pothos.
Pri., Prinos.
Pro., Propylaminum.
Pru. p., Prunus padus.
Pru. s., Prunus spinosa.
Pso., Psorinum.
Pte., Ptelea trifoliata.
Pul., Pulsatilla.
Pul. n., Pulsatilla nuttalliana.
Pyrth., Pyrethrum.
Pyrus, Pyrus.

Qua., Quassia.

Ran. a., Ranunculus acris.
Ran. b., Ranunculus bulbosus.
Ran. g., Ranunculus glacialis.
Ran. r., Ranunculus repens.
Ran. s., Ranunculus scelcratus.
Rap., Raphanus.
Rat., Ratanhia.
Rei., Reinerz.
Res., Resina itu.
Rha. c., Rhamnus catharticus.
Rha. f., Rhamnus frangula.
Rhe., Rheum.
Rhodi., Rhodium.
Rhodo., Rhododendron.
Rhu. g., Rhus glabra.
Rhu. t., Rhus tox.
Rhu. v., Rhus venen.
Ric., Ricinus.
Rob., Robinia.
Ros., Rosmarinus.
Rum. a., Rumex acetosa.
Rum. c., Rumex crispus.
Rus., Russula.
Rut., Ruta.

Saba., Sabadilla.
Sabi., Sabina.
Sac., Saccharum album.
Saln., Salicinum.
Sal. ac., Salicylicum acidum.
Sal. n., Salix niger.
Sal. p., Salix purpurea.
Sam., Sambucus.
Sam. c., Sambucus canadensis.
Sang., Sanguinaria.
Sant., Santoninum.
Sap., Saponinum.
Sarr., Sarracenia.
Sars., Sarsaparilla.
Sca., Scammonium.
Sch., Schinus.
Scol., Scolopendra.
Scor., Scorpio.
Scr., Scrophularia.
Scu., Scutellaria.
Sec. c., Secale cornutum.
Sed., Sedinha.
Sel., Selenium.
Senec., Senecio.
Seneg., Senega.
Senn., Senna.
Sep., Sepia.
Ser., Serpentaria.
Sil., Silicea.
Sin. a., Sinapis alba.
Sin. n., Sinapis nigra.
Siu., Sium.
Solnm., Solaninum.
Sol. a., Solanum arrebinta.

Sol. m., Solanum mammosum.
Sol. n., Solanum nigrum.
Sol. o., Solanum oleraceum.
Sol. p., Solanum pseudo-capsicum.
Sol. t., Solanum tuberosum.
Sol. t. æ., Solanum tuberosum ægrotans.
Spig., Spigelia.
Spig. m., Spigelia marilandica.
Spigg., Spiggurus.
Spira., Spiranthes.
Spire., Spirea ulmaria.
Spo., Spongia.
Squ., Squilla.
Stach., Stachys betonica.
Stan., Stannum.
Stap., Staphisagria.
Stic., Sticta pulmonaria.
Stil., Stillingia sylvatica.
Stram., Stramonium.
Stro., Strontium.
Stry., Strychninum.
Sul., Sulfur.
Sul. h., Sulphuretted hydrogen.
Sul. i., Sulfur iodatum.
Sul. ac., Sulfuricum acidum.
Sulphs. ac., Sulphurous acidum.
Sum., Sumbul.
Syph., Syphilinum.

Tab., Tabacum.
Tana., Tanacetum.
Tang., Tanghinia.
Tann., Tannin.
Tarax., Taraxacum.
Tarent., Tarentula.
Tart. ac., Tartaric acid.
Tax., Taxus baccata.
Tel., Tellurium.
Tep., Teplitz.
Ter., Terebinthina.
Tet., Tetradymite.
Thal., Thallium.
Thea, Thea.
Ther., Theridion.
Thev., Thevetia.
Thu., Thuja.
Til., Tilia.
Tit., Titaninm.
Ton., Tongo.
Toxa., Toxicophlœa.
Toxs., Toxicophis.
Trac., Trachinus.
Trad, Tradescantia.
Trif. p., Trifolium prætense.
Trif. r., Trifolium repens.
Tril., Trillium cernuum.
Trim., Trimethylaminum.
Trio., Triosteum.
Trom., Trombidium mus. domes.
Tus. f., Tussilago fragrans.
Tus. p., Tussilago petasites.

Upa., Upas.
Ura., Uranium nitricum.
Ure., Urea.
Urt. c., Urtica crenulata.
Urt. g., Urtica gigas.
Urt. u., Urtica urens.
Ust., Ustilago.
Uva, Uva ursi.

Vac., Vaccininum.
Val., Valeriana.
Verat., Veratrum album.
Verat. v., Veratrum viride.
Veratn., Veratrinum.
Verb., Verbascum.
Vesp., Vespa.
Vich., Vichy.
Vin., Vinca.
Vio. o., Viola odorata.
Vio. t., Viola tricolor.
Vip., Vipera.
Vip. l. f., Vipera lachesis fel.
Visc., Viscum album.
Vös., Vöslau.

Weis., Weisbaden.
Wild., Wildbad.
Wye., Wyethia.

Xan., Xanthoxylum.

Yuc., Yucca.

Zing., Zingiber.
Ziz., Zizia.
Zn., Zincum.
Zn. a., Zincum aceticum.
Zn. c., Zincum cyanatum.
Zn. f., Zincum ferrocyanatum.
Zn. m., Zincum muriaticum.
Zn. p., Zincum phosphoratum.
Zn. s., Zincum sulfuricum.

EAR: GENERAL.

[NOTE. — As has been said in the preface, this repertory is not my ideal, but may prove a stepping-stone to a better. In the classification of symptoms under anatomical subdivisions, the attempt has been made to place symptoms in their proper physiological relations. This may be a matter of uncertainty in some instances, on account of faulty expression, or lack of knowledge of exact functions, as now understood. This effort may not commend itself to those who care only for symptoms, but it will be appreciated by those who labor to link symptom and function.

The author will be pleased to have his colleagues report any corrections or additions, that they may be noted for future reference. The emphasis placed upon remedies is marked by three degrees, — in small type, Italics, and capitals.]

Ear, ACHING. All. c., Aloe, Anac., Arn., *Asaf.*, Asaf. (r.), Asar., Astac., Aur., Bar. c., Bell., Bell. (l.), Bor., Bism., Bro. (r.), Bry., Calc. cau., Cannab. i., Canth., *Cap.*, Caus., Cham., Cimic., Cinch., Cle., Coc. c., Coloc., Con., Cup., Dig., Ery. a. (l.), Euphr., Fl. ac., For., Grap., Guai. (l.), Ham., Hell., Hyos., Ig., Indg., Iod., Ip., Jat., Jug. (r.), K. ca., Lach., Lact., Lau., Lith. (l.), Lob. i., Lyc., Mag. c., Mang., Mang. (l.), Merc., Merc. i. r., Merl., Mez., Mez. (l.), Mur. ac., Na. m. (l.), Nx. m., Nx. v., Olnd., Osm., Petrol., Pho., Pru. r., Pso. (r.), Ran., Rhu. t., Rum. c., Rut., Saba., Sabi., *Sars.*, Seneg., Sep., Sil., SPIG., Spo., STAN., Stap., Sul., Sul. (l.), Sul. ac., Tab., Tarent., Thu., Ust., Verat., Verb., Vio. o.; MORNING, in bed, Merc. i. r.; on rising, Fer.; > taking wax from ear, Calc. s. (l.); AFTERNOON, Euph. (r.); EVENING, Berb. (r.), K. bi., Lyc. (l.), Na. m.; 1 P.M., Mit. (r.); 2 P.M., Pau. p.; 5 P.M., Ham. (r.); from being in open air, Sep.; after MIDNIGHT, Sep.; after walking in wind, Sep. (r.); NIGHT, *Dul.*; in OPEN AIR, Euphm., Lyc.; on BLOWING NOSE, Sil. (l.); before BREAKFAST, Stach.; at every COUGH, *Cap.*; after LYING down, Sang.; during MENSES, Aloe; (AMEL.), pressure, Ham. (r.); entering warm room, 5 P.M., Sep.; removing wax, Calc. s. (l.); WITH hard hearing, Ip.; shooting, Lob. i. (r.); ACUTE, Olnd.; as if BURNT, extending outward into ear, Bry. (l.); in CHILDREN, ALL. C., *Bell.*, CHAM., *Dig.*, *Merc.*, PUL., *Rhu. t.*, ZN. (boys?); DRAWING, on approach of thunderstorm, or cloudy, windy weather, Rhodo.; EXTENDING to finger-ends, Ham.; to jaw, on least dampness, Itu.; from l. to r., Calc. ph.; to malar bone and molar (*Spig.*), (r.); into parotid gland and mastoid process, when walking in wind, > warmth, 5 P.M., Sep.; outward, walking in cold wind, 4.30 P.M., Sep.; into teeth, Hydrphb., Xan. (r.).

AIR in, sensation of. Mez. (r., then l.).

BAR through jaws from ear to ear, sensation of. Hur.

BEATING. See PULSATION.

BITING. Pso. (l.); like electric sparks here and there, 8 to 10 P.M., Phe.

BLOOD RUSHING into. Gel., Hydrphb, *Lyc.*, Pul.; sensation of hot, Lyc.

BLOW, pain as from. Ail. (r.), Anac., Arn., Bell., Con., Nx. v., Ol. an., Rut., Spig., Verat.; extending inward, Bell.; outward, Con.

BLOWING, pulsative, at night. Sep. (r.).

BOARD before, sensation of. Arg. (l.).

BOIL. K. ca.; sore, Spo. (l.).

BORING. Alum, Am. car., Aur., *Bar. c.*, Bar. c. (r.), BELL., Berb., Cannab. i. (r.), Canth. (l.), Carb. a., Carb. a. (r.), Carbn. s. (r.), Caus., Caus. (r.), Cup., Cup. ac. (r.), Euphr., *Hell.*, *Indg.*, K. ca., K. iod., Lau., Lepi., Mag. c., Mag. c. (r.), Mag. m., Mag. m. (r.), Mag. s. (l.), Merc. i. f., Mil., Na. m., Ol. an., Phe., Plat., Pb., Ran. s., Rhodo., Saba., SIL., *Spig.*, Stan., Stan. (r.), Sul., Sul. (< r.), Thu.; MORNING, Alum.; FORENOON, Mag. c. (l.); AFTERNOON, Alum.; AFTER DINNER, Pb. (r.); on PRESSURE, Alum.; WITH sticking, Mag. m.; ticking, Nicc (r.); boring behind ear, Cup.; ACUTE, Merc. i. f. (l.); has to scream, Bar. c.; as if something wanted to COME OUT, Am. m., Iod.; DEEP-SEATED, Aur. m. n., Merc. i. f. (l.), Phe.; EXTENDING inward, Am. m., Mag. c.; with stitches, Mag. c.; to jaw, Phe.; outward, Am. m. (r.), Aur. m., Merc. i. f., Merc. i. f. (l.); towards parietal bone and temple in afternoon, Indg.; OBTUSE, Mag. c.

BRUISED feeling. Am. car., Stry.

BURNING. Acon., Æs. h., Aga. m., Aloe, Alum., Aml. n., Am. m., Angu., Ant. cr., Ant. cr. (l.), Arn., Arum t. (r.), Ars., Asaf., Aur., Aur. s., Bell, Berb., Bro., *Bry.*, *Calc. ostr.*, Camph., Cannab. i., Carb. a., Carb. v., Casc., Caus., Chel., Chin., Cle., Cyc. (r.), Daph., Dig., Dro., Dro. (r.), Fago. (l.), Grat., *Ig.*, Jat., Kre., Lau., Lil. s., Lob. s., Lyc., Lycps. (r.), Mag. c., Mag. c. (r.), Mag. m., Merc., Merc. (r.), Merc. sol., Naj., Na. m., Nitr., Olnd., Ol. an., Pho. ac., Pic. ac., Plat., PUL., Ran. b., *Rhu. t.*, Saba., Saba. (r.), Sabi., Sang., Sars., Spig., Spire., Spo., Stap., Stry., Tab., Tarax., Tel., TEL. (l.), Ter., Til., Zn.; NOON, Stry.; AFTERNOON, Stry. (l.); EVENING, Bro. (l.), Ham. (r.), Zn.; 1.40 P.M., Equ. (r.); 2.30 P.M., Jug. (r.); at NIGHT, Stry.; before going to sleep, Stil. (l.); after RUBBING, Grat. (r.); on TOUCH, Cop. (l.); from DRY WARMTH, Bry.; while YAWNING, < open air, in evening, > after eating, Acon.; > PRESSURE, Ham. (r.); WITH purulent bloody discharge, Pul.; otitis, Kre., Nitr.; heat in head, Jat.; heat in face after a chill, Tarent.; red cheeks, Sang.; hot stage, Ran. b.; sweat, Acon. (l.); FOLLOWED by hard hearing, Cap.; as from a HOT COAL, Bro.; as from coming from COLD into a warm room, K. n.; DEEP-SEATED, Merc. sol. (< l.); like ELECTRIC SPARKS, 8 to 10 P.M., Phe.; EXTENDING inward, Am. m.: down angle of jaw, Bov. (r.); outward, Aur. m., Sep.; 5.30 P.M. on walking in open air, Am. m. (l.); like fire, Mag. c.; as if FROST-BITTEN, Aga. m.; INTERNALLY, as from a red-hot coal, Tep. (r.); PAINFUL, with purulent and bloody discharge, Chin., Sul.

BURROWING sensation, as of ANIMALS, < lying still, Ant. cr.; in ear on which he lies at night as if SOMETHING WOULD COME out, Am. m. (r.); CONGESTIVE, extending to lower jaw, Am. c. (l.); EXTENDING to side of neck, clavicular region, last back teeth, and side of occiput, Coc. c. (l.).

BURSTING sensation. Aml. n., Calc. cau. (l.), Caus., Lyc., Nit. ac., Pho.; with each beat of the heart, Aml. n.; as of a leaflet, in evening, Gam.

CATARRH. Bar. m.

CLOSED sensation. Lachn., Merc. i. f. (r.), Nit. d. s. (l.), Pul. n. (l.), Spig.; MORNING, Bro. (r.); in OPEN AIR from wind blowing, Spig; WITH stitches, Glo.; ALTERNATELY, as if closed and deaf, Cocc.; as if DEAF, Nit. ac.; though hearing well, Merc. i. f. (r.); OPENING with a dull (dead) sound, Mar.; SUDDEN, Tanac.

COLDNESS. Amyg., Ars., Berb. (l.), Calc. c., Calc. ph., Chel. (r.), Cic. v., Dul., K. ca., Lach., Lyc., Mang., Mang. (r.), Meny., Merc., Plat., Pso. (r.), Seneg., Stan., Stap., Tel., Ter. (l.), Verat. v.; EVENING, Mez.· 5 P.M., Paeo.; in WARM BED, Merc. sol.; WITH bubbling sound, Berb.; numbness extending to cheeks and lips, Plat.; pulsation, Berb.; twitchings, Plat.; ALTERNATING with heat, Verat. (r.), of external ear, Verat.; as after TAKING COLD, Wild.; EXTENDING through ear, Seneg. (l.); as if WATER had got in, Meny.

COMPRESSION. Sars.

CONGESTION. Aloe, Coloc., Fl. ac., Glo., Hyos., Nx. m., Pul.; with gastric symptoms, Tarent.

CONSTRICTION. Thu. (r.); pain as if constricted internally, Dig.

CONTRACTION. *Anac.*, Asar., Bry., Caus., Croc., Dig., Dro., Lach., Merc., *Sars.*, Sars. (r.), Spig., Spo., Thu.; WITH shocks, Bell.; SPASMODIC, afternoon, when sitting, Æth. (r.); SENSATION of, evening, after lying down, < sleeping on that side, Caus. (l.).

CRAMP. Aga. m. (l.), Aloe, Ambra, *Anac.*, Angu., *Arn.*, Ars., Asar., *Bell.*, Calc. cau., Calc. ostr., Carb. a., Carb. v., Caus., *Cham.*, Chin., *Cina*, *Cle.*, Coc. c., Colch., Coloc., Con., CROC., Crot. t., Dig., Dro., *Dul.*, Euphm., Gran., Grap., Hell., Iod., K. ca., K. n., Kre., Lyc., Mang., Mar., Meny., MERC., *Merc. sol.*, *Mur. ac.* (l.), Na. c., Na. m. (l.), Nit. ac., *Nx. m.*, Nx. v., Olnd., Par., Petrol., Pho., *Pho. ac.*, Plat., PUL., Ran. b., Ran. s., Rhe., Rhodo., Rhu. t., Saba., *Sam.* (r.), Sep., Sil., SPIG., Spo., Stan. (r.), Stram, Stap., *Sul.*, Thu., Verat., Verben., Zn.; EVENING, Thu. (r.); on WALKING in open air, Spo. (l.); FOLLOWED by stitches, Thu.; INTERNALLY and EXTERNALLY, in evening, Ran. b. (l.).

CRAWLING. Ars., Cinch., Colch., Dul., Grat., Lau. (l.), Mang., Meny. (l.), Mep. (l.), Merc., Spo., Sul., Sul. ac.; WITH tearing, Zn.; DEEP-SEATED, Lau.; DISTURBING night's rest, Nx. v.; PAINFUL, like digging, going after a while to lower jaw, Am. car. (l.).

CREEPING, with tearing. Sul. ac., Ton.

CUTTING. Arg., Bell., Canth., Coloc., Mur. ac., Petrol., Petrol. (l.), Sil.; extending inward, Agn., Bell.; outward, Astac.

DARTING. Glo., K. iod. (r.), Kalm., Lepi., Rhodo. (l.); while yawning, < evening or in open air, > after eating, Acon.; acute, Stry.; extending to shoulder, Lepi.

DIGGING. Am. car., Am. m., Ant. cr., Colch. (r.), Gel., Gel. (r.), Hell., Plat.; AFTERNOON, Gel. (r.), Merc. (l); WITH hard hearing, Jac.; DEEP in, Stry. (l.); EXTENDING to nostril, with noise and hard hearing, Jac.; to lower jaw, Am. car.; as from INSECTS, K. iod.

DISTENDED sensation. Bell., K. iod., Lau. (r.), Nit. ac., Pul.

DIVING, sensation as when, while writing. Rap.

DRAGGING, downward, Berb. (r.); outward, Caus., Euphr.; 4 P.M., Canc. f.; pulsating, > stooping, Cannab. s.

DRAWING (including PULLING). Acon., Angu. (r., then l.), Aur. m. (r.), *Bar. c.*, Bell., Berb., Bov., Bry. (r.), Calc. c., Carb. a., *Caus.* (r.), Chel. (l.), Coc. c., Con. (l.), Crot. h., K. ca., Lyc., Mag. m., Merc. sol., Mez. (l.), Mil. (l.), Mos. (r.), Na. m. (r.), Nicc., Nit. ac. (r.), Op., Petrol., Pul. n., Rhodo., Sep. (r.), Sil., Stan. (l.), Stap., Sul. (r.), *Verb.* (l.); at NIGHT, Bar. c.; WITH stopped feeling, Lach.; from taking COLD, Glo. (r.); after LYING down, Sul. (r.); on MOTION OF LOWER JAW, Stan. (r.); on PRESSING posterior part, Rap.; > SNEEZING, Mag. m.; ACUTE, Pho.; ALTERNATELY, Lact.; as from taking COLD, Til (l.); CONTRACTIVE, Croc. (r.); CRAMPY, Crot. c. (r.), Petrol., Sil. (r.), Val. (l.); DEEP-SEATED, Ost.; DULL, extending inward, Dro. (r.); EXTENDING to forehead, Dign. (l.); inward, Dro., Nx. v., *Verb.* (l.); towards malar bone, Spig. (l.); into neck, Bell., Sil. (r.); to shoulder-joint, Na. m.; outward, Pul. n.; to teeth, Mos. (r.); Chel. (l.); to vertex, Arn. (l.), forenoon, Gam. (r.); to zygoma, Hype. (l.); sensation as if something were pulled from shoulder, Lepi.; INTERNALLY, Spo. (r.), and EXTERNALLY, Stan. (r.); SPASMODIC, Pho. ac.; STUPEFYING, Asar.; TEARING, Plat.; TWITCHING, Plat. (l.).

DULNESS. Chel., Chlo. (l.).

ENVELOPED or stopped, sensation as if. Bor.

ERUPTION, crusty, and on cheek. Elap.

FALLING to and fro, something seems, with clucking, Grap., and bursting, then singing, Saba.; sensation of something falling on the floor, and cracking, with, and followed by, rushing, Saba.

FANNING before, sensation of. Calc. c. (l.).

FLAPPING. Caic. c.; like butterflies' wings, 8 A.M., Jac.

FORCING. Spo.; as of brain through skull, Na. m.; as if something were forcing out, Merc. sol., PUL.; towards ears, Lyc.

FORMICATION. Ambra, Bar. c. (l.), Calc. c. (r.), Caus. (l.), Coloc., Cop., Der., Dro. (l.), Grat., K. ca., Lau., Merc. sol. (l.), Mil. (l.), Nx. v., Plat., Physo., Rat. (r.), Sam., Sep. (r.), Sul. ac. (l.), Tong. (r.); in MORNING, Zn.; while EATING, Lachn.; EXTENDING to lower jaw, Am. c. (l.); FINE, Na. m. (r.); ITCHING, Spig. (r.); TICKLING, Chin.

FULNESS. Æs. h., Berb. (r.), Bon., CANNAB. I., Dig., Eup. pur., Iod., Fer., Glo., Jug. (r.), Na. ph., Na. slfc., Pen., Physo., Stry., Verat. v. (l.); MORNING, Thu. (r.); 8 A.M., Ham. (l.); AFTERNOON, Stry.; EVENING, Na. ph.; after STITCHES leave it, Iod.; < EXCITEMENT, Dig.; WITH pain, Berb.; fulness around ears, Glo.

GNAWING. Arg., Dro., Indg., K. ca. (l.), K. iod. (l.), Mur. ac., PHO., Sul., Sul. (l.), Tab., Tab. (r.); afternoon, > rubbing, Indg. (l., then r.); evening, Mur. ac. (r.).

GRIPING. At night on waking, Carbon. s. (l.).

GURGLING. > boring with finger, Castor. (r.); with increased earwax, K. ca.

HEAT. Acon., Aloe, Alum., Aml. n. (l.), Angu., Arn., Ars., Asaf., Asar., Aur. m. n., Bell., Bon., Bro. (r.), Bry., Calc. c., Calc. astr., Calc. ph., Canna, Canth., *Carl.*, Casc., Caus., Cham., Cle., Cic. v., Coloc. (r.), Coc. c. (l.), Crot. h.,

Fago. (l.), Gas., Grap., Hep., Ign., Iod., Jac., Jat., K. ca., Kis., K. n., Kre., Lach., Lyc., Mag. m. (r.), Manc., Mang. (l.), Meny., Merc., Mur. ac., Na. m., Nit. ac., Nit. ox., Nx. v., Ol. an., Oena., Paeo., Pau. p., Par., Peti., Petrol., Pho. ac., Pho., Plat., Pso. (r.), *Pul.*, Rap., Rhodo., Sabi., Seneg. (r.), Sep., Sep. (r.), Sil., Spo., Sul. ac. (r.), Sul., Tab., Tarent. (r.), Ter. (r.), Thu. (r.), Til. (r.), Verat., Wild. (r.), Zn., Zing.; at 10 A.M., Pip. m.; AFTERNOON, Cannab. s.; after drinking coffee, Na. m. (r.); EVENING, *Alum.*, *Cap.*, *Carb. v.* (l.), Na. m. (l.), Na. n., Sabi.; 4 P.M., Ped.; 10 P.M., Stry. (l.); MIDNIGHT, > going to sleep, Alumn.; NIGHT, before falling asleep, Sul.; when LYING DOWN, Ars., Hype.; WITH burning and hard hearing, Jac.; itching, Nit. ac.; otitis, K. ca. (r.); painful sensations, Petrol.; drawing pain, Sep.; digging, burning pain extending to left nostril, Jac.; pressure in occiput, Gran.; chill, Pul., in back, Asaf., in upper arms, *Ig.*; coldness of body, Acon.; pleasant warmth, Mag. m.; ALTERNATING with coldness, Verat. (r.); drawing, Lach.; like HOT BLOOD, Calc. c.; EXTENDING from ears, Olnd., Sep.; over half of head, Chel. (l.); to angle of jaw, Bov. (r.); over occiput to nape of neck, Spo.; to pharynx, evening while riding, Nx. m.; EXTERNAL, with twitching, Pul.; INTERNAL, Arn.; as if near a stove, Ant. t. (l.).

HEAVINESS. Gel., Glo.; before menses, Crot. h.; in and before ears, with a stopped sensation, Carb. v., with hard hearing, CARB. V.

HOLLOW sensation in morning, > after dinner, Nx. v.

ICE within, sensation of, in warm bed. Merc. sol.

INFLAMMATION. ACON., Aloe, Ant. cr., *Bell.*, Bor., *Benzn.*, *Cact.*, CALC OSTR., Cannab. i., *Canth.*, Carb. v., CHAM., Cist., *Cocc.*, Coloc., *Dul.*, Fer., Fl. ac., Gel., Glo., HEP., K. CA., Kre., LYC., Manc., MAG. C., Mag. m., *Merc.*, MERC. S., Na. m., Nitr., *Nx. v.*, Par., *Pho.*, PHO. AC., Pte. (r.), PUL., Pul. n., RHU. T., Rhu. v., Sang., Sep., Sil., Spo., Sul., *Tel.*, Thu., Ulmus c., Zn.; WITH anguish, Bell.; burning, Kre., Nitr.; constant buzzing, Merc.; caries of ossicula, Lyc., Sil., Sul.; great distress from noise, Merc.; partial deafness, Merc.; discharge of pus, Bell.; increased hardness of hearing, Thu.; heat, K. ca. (r.); itching, Nitr.; pain, Bell.; agonizing pain when going to bed, lasting until morning, Merc.; ulcerative pain, Mag. c.; violent pain, Pul.; pyæmia, Pho.; redness, Kre., Mag. c; sensitiveness, Mag. c., intense, and of adjoining parts, Merc.; soreness in meatus, Merc.; jerking, starting, Nitr.; stitches, Alum. (l.), Merc., Merc. c.; suppuration in meatus externus, Kin.; swelling, Kre., Nitr.; swelling in meatus, Calc. ostr., Cist., K. ca., Na. m., Sep., Tel., Thu., Zn. a., and purulent discharge, Pul.; swelling around ear, Pul.; painful and considerable swelling of mastoid process, Merc.; tearing, Merc.; delirium, Bell., Pul.; tensive pain in head, Bell.; sometimes paleness, sometimes flushed face, Kalm.; great thirst, Pul.; vomiting, Bell.; costiveness, Pul.; aphonia, Bell., Rhu. t.; fainting-fits, Bell.; coldness of limbs, Bell.; PERIODICAL, Rhu. v.

ITCHING. Acon., Ambra, *Aga. m.*, Alum., Am. m., *Anac.*, Apis (r., then l.), Aq. p., ARG., Arg. (l.), *Bar. c.*, Bell., Benz. ac. (l.), Berb., Brach., Bov., Bry., Calc. c. (l.), Calc. ostr., Calc. p., CAP., Carb. a. (r.), Carb. v., *Carl.*, Caus. (l.), Cent., Chel. (r.), Cinnb., Coc. c. (l.), Con., Cup., Cyc., Elap., Fago. (l.), Fl. ac. (r., then l.), For. (l.), Grap., Grat., Ham. (l.), *Hep.*, IG., *K. ca.*, K. iod., K. iod. (l.), K. n., Kre., Lach., Lachn., Lau., Lau. (r. and l.), Lep., Lip., Lyc., Mang. (l.),

Meny. (r.), Mep. (r.), Merc. i. r. (r.), Merc. sol., Mez. (r.), Mil. (r.), Mos., Mur. ac. (r. and l.), Na. m. (r.), Na. slfc. (l.), Nit. ac., Nitr., Nx. v., Ol. an., Paeo., Ped., Ped. (r.), Petrol., Petrol. (l.), Phel., Phel. (r.), PHO., Pho. ac., Pso. (r.), Pul. (r.), Rat. (r.), Rhe., Rhodo., Rhodo. (l.), RHU. T. (l.), Rum. c. (r.), Sam., SARS., Sars. (l.), Sed. (l.), Sel., *Sep.*, Sil. (< l.), Spig., Stan. (l.), Stry., *Sul.* (l.), Sul. i., Tarax., Tarent. (r.), TEL. (l.), Ton. (r.), Verat. v. (l.), Vio. t., Wies., Wyeth., Zn. (l.); in MORNING, Na. c., Mag. m.; AFTERNOON, Aga. m. (l.), *Arg.*, Pul.; EVENING, Acon., Arg., Cala. (r.), Dios., Grap. (l.), Ol. an., Ped. (l.), Pte. (r.), Pul.; when walking, Bor. (l.); 2.30 P.M., Grat. (r.); 3 P.M., Lau., 4 P.M., Ol. an., 5 P.M., Fago. (l.); 6 P.M., Pte. (r.); 8 P.M., Ol. an., Rum. c. (l.); NIGHT, *Merc. i. f.*, Sep. (r.), Stry.; after oozing of LYMPH, Grap.; on RISING, Trom.; while YAWNING, Acon.; (AGG.), evening, Acon., Na. m.; in open air, Acon.; on gasping, Acon.; rubbing, Alum., scratching, Ton.; stooping, Lepi.; going from cold air into warm room, Coc. c.; (AMEL.), rubbing, Mez., Phe.; scratching, Am. m., Caus., Chel., Lau., Na. slfc. (r.); touch, Na. m.; WITH buzzing and roaring, Sep.; discharge, Anac., Bar., Merc., Petrol., Sep.; fetid discharge and deafness, Bov.; eruption and discharge, Sul.; dull hearing, Bov., and suppuration, Am. car., Bov.; heat, Nit. ac.; inflammation, Nitr.; sensation like the jumping of fleas, Zn.; pimples on ear, Berb.; pulling of ear, Pul., or like a worm, Rhodo.; roaring, Sep.; swelling and discharge, Tel.; swelling of lobe, and offensive discharge, Tel.; warmth, Berb.; much earwax, Sed. (l.); in forenoon, Cyc., Mur. ac., Sep.; running of wax, Am. m.; accumulation of white wax, Sep.; itching of cheek, Elap.; irritation and red, bleeding blotches in throat, Fl. ac.; FOLLOWED by humming, Na. m.; sore pain, Arg.; BITING, Caus. (l.), Berb., Verat.; crawling, Verat.; BORING, Lau., Mil.; crawling, Mil., Ton.; BURNING, AGA. M., Berb., Calc. c., Calc. ostr., Com. (r.); CORRODING, arg.; crawling., Plat.; CRAWLING, Alum., Ambra, *Am. car.*, Ant. cr., BAR. C., Calc. ostr., Carb. v., CAUS., Coc. c., Colch., Dro., Lachn., LAU., Merc., Mil., Nx. v., Phe., Plat., Rhe., Rhodo., RHU. T., Sam., SEP., Spig., Sul., Ton.; after dinner, > boring in with finger, Phe.; crawling and creeping, Nx. v.; crawling and tickling, Mang.; as if something alive were in ear, Rhu. t.; painful crawling, Am. car., Bar. c., Calc. ostr., Sul.; as after sleep, Lau.; DAILY, Sep.; DEEP-SEATED, *Cap.*, Colch.; in left ear, followed by yawning, later itching in right ear, Rum. c.; crawling, Carb. v., Caus., Petrol.; tickling, Mang., Pul.; extending from ear to ear, under chin, Lepi.; to neck, Coc. c.; FINE, Pul.; HOT, tickling, Lach.; INTOLERABLE, *Coc. c.*, Fl. ac.; PAINFUL, Pso.; FORCING SCRATCHING, Mos.; till bloody, Arg.; STICKING, Berb., Lach., Pho. ac., Spig.; fine, Berb., Lau.; TEARING, Rhe.; TICKLING, Acon., AGA. M., Ambra, Am. m., Ars., Bov., Calc. ostr., Chel., Crot. c., *Cup.*, K. CA., LYC., Mag. c., Mag. m., Mang., Na. c., Petrol., Rhodo., Rut., Sam., Zn., > scratching, Bro.; pleasant, Pul.; VIOLENT, *Arg.*, Calc. ostr., Carb. v., Con., Nitr., *Pho.*, Sep., *Sul.*, Ton.; VOLUPTUOUS, tickling, Ars., Rum. c.

JERKING. See TWITCHING.

JUMPING of fleas, sensation of. Mos. (l.), Spig.; on putting finger in, Zn. (l.); with itching, Zn.; with waving, Mos.; jumping as from swashing of water, when stepping suddenly, Spig.

LACERATION. See TEARING.

LEAF lay in front of the ear, feeling as if, without affecting the hearing. K. iod.

LIVING were in, sensation as if something. Rhu. t.

LUMP in, dream that there is a. Cinnb. (r.).

MEMBRANE before, sensation of. Cannab. s., Verat.; > shaking head, and boring finger into ear, Sel.

MIST before, sensation of. Par.

NUMBNESS. Calc. i., Manc., Thu. (l.), *Verb.* (l.).

OBSTRUCTION. See STOPPED SENSATION.

PAIN. *Acon.* (l.), (r.), Æs. h., Æth., *Aga. m.*, Agn., *All. c.*, *Alum.*, AMBRA, AM. M., *Anac.*, ANGU., Apis (r.), Ap. v., Aral., ARN., ARS., ASAF., Atro., Bap., *Bell.*, Benz. ac. (r.), Berb., Bor., Brach., Bro. (r.), Bry., Cact., Cala., Calc. c. (r.), CALC. OSTR., Canch., CANNAB. I., Cannab. i. (l.), CAP., CARB. V., Carbn. s. (l.), Casc., CHAM., Chel., CINA, CINCH., Cinnb., Cit. v. (r.), CLE., Cob. (l.), Coch., COLCH., Coloc., CON., CROC., Crot. t., CUP., Cur. (r.), CYC., Der. (< l.), *Dig.*, Dios. (l.), DRO., DUL., Ery. a., Euphm., Euphr. (r.), Fl. ac. (r.), For. (l.), Gel., *Grap.*, GUAI., HÆM., Hæm. (r.), HELL., Hur. (r.), Hydrs. (r.), Kalm., K. bi., K. ca., K. iod., K. n., LACH., Lach. (r.), Lepi., Lil. s. (l.), Lith., Lob. s., LYC., Mang., Ment. p., Meny, MERC., Merc. i. f. (r.), *Mez.*, Mil. (r.), Mit., Mor. (l.), MUR. AC., Muru. (l.), Na. c. (l.), Na. m., NX. M., Nx. m. (r.), OLND., Op., Osm., PAR., Pau. p., PAU. S., Petrol., PHO., *Pho. ac.*, Phyt. (< r.), Plan., PB., Pb. (l.), Pru. s., *Pso.*, Pte. (r.), *Pul.*, Pul. n., Rat. (r.), RHE., Rum. c., SABA., SAM., Sang. (l.), Sarr., SARS., SIL., Sil. (r.), SPIG., SPO., STAN., Stic., Stram., SUL., SUL. AC., Tarent. (r.), Tel., Ton., VERB., VIO. O., Vio. t., Wies., Zn.; in MORNING, Rum. c. (r.); in bed, Carb. v., Nx. v.; till evening, Tarent. (r.); FORENOON, Sul.; 8 A.M., from cough, Dios.; 9 A.M., after going out, Tel. (l.); 11 A.M., Dios. (r.); AFTERNOON, Rum. c., Sul. (r.); on blowing nose, Dios. (l.); EVENING, Alum., Brach. (l.), Cob. (l.), Colch., Dios. (r.), Hyos. (r.), Ran. b., Sep., Sul. (r.); while sitting, Lach.; 1 to 2 P.M., Chin. s. (r.); 3.30 P.M., Am. car.; at NIGHT, Lach., Nitr., Sep.; 10 P.M., till morning, Spig.; in OPEN AIR, Con.; from CHEWING, Apis (l.), and talking, Nx. v.; from taking COLD, > heat in hot rooms, Bell.; when CLEANING, Physo., Sul.; at every stroke of a HAMMER, Sang. (r.); from MUSIC, Tab.; from NOISE, Gad. (r.); PUTTING FINGER in, Zn. o.; on RAISING HIS FACE, Nx. v.; TALKING and chewing, Nx. v.; from TOBACCO, Rap.; on TOUCH, Chin. (l.); on TURNING EYE OUTWARD, Rap.; WALKING, Mang., Sul.; entering WARM ROOM from cold, raw air, not the reverse, Na. slfc.; while WRITING, Physo.; (AGG.), cold air, Bry.; lying, Bell., on painful side, Bell.; coming near the range, Coch.; after entering a room, or being in bed, Nx. v.; stooping, Coch.; talking, Mag. m., Mang., Nx. v.; touch, Sul.; (AMEL.), boring in with finger, Sul.; on rising, Carb. v.; stooping, re-appearing on raising head, Cannab. s.; warmth, Lach., or wrapping up, Mur. ac.

—WITH the painful sensations, fulness in EARS, Berb.; hard hearing, *Asar.*, and chilliness, *Bell.*, *Cham.*, *Merc.*, PUL.; lost hearing, Bry., Cyc.; heat, Petrol.; inflammation, Bell.; pulsation, at night, Rhu. t.; roaring, Ars.; swelling of external ear, and headache, Pso.; tingling, Osm.; ILL-HUMOR, Cham.; HEADACHE, Berb., Ham., Nitr.; pressing or gnawing pain, Ran. s. (r.); pain in forehead, Sang.; pressing in forehead, Nit.

ac.; pain in temples, Hyos.; confusion, Hyos.; stitches in corresponding side of head, K. bi. (l.), Kalm. (r.); soreness of scalp, Lach.; FACE-ACHE, Lith.; pain in cheeks, Pho. ac.; redness of cheeks, Sang.; tickling on right cheek, Bro.; drawing in joint of jaw, Na. c.; lameness of jaw, Lach.; pain in jaw, then pricking in ear, Hur.; TOOTHACHE, Plan.; drawing pain in teeth, Pho. ac., Ran. s. (r.); gnawing pain in upper teeth, K. iod.; frequent SPITTING, Hæm.; burning in THROAT, Hæm.; pain in STOMACH, K. ca.; great nausea, Dul.; anxiety in ABDOMEN, Aloe, Glo.; much MICTURITION, Thu.; anxiety in CHEST, K. ca.; swelling, and painfulness to touch, of a gland in the NECK, K. bi.; pain in LEGS, Kalm.; gnawing in hollow of knee, K. iod.; RUNNING like a crawl over the same side, K. iod.; FEVER stage, Acon., *Apis*, *Cala*, Calc. ostr., Dig., Nx. v., *Pul.*, Sul.; cold feet, Stan., Thu.; sweat, ACON., Bell., COLOC., Caus., *Lyc.*, *Merc.*, Na. m., Nit. ac., *Pul.*, Sep., Sul., Thu.

PAIN, FOLLOWED by hard hearing, Nit. ac.; pinching, Nitr.; stitches, Berb.

—ACUTE, Bell., Der., For. (l.), Plumbg. (l.), Physo. (r.), Pte. (l.), Pul., Pul. n. (r.); FORENOON, Fago. (r.); 8 A.M., on turning suddenly to right or left, Erio. (r.); 9 P.M., Dios. (l.); all NIGHT, Merc.; on CHEWING, Op. (l.); INCLINING head, > evening holding it up, Chin. s.; when riding in cold WIND. Ars. i. (< l.); WITH great swelling of ear, and pain in the head as if it would set him crazy, Pso.; EXTENDING inward, Tel.; making him start, Sul. ac.

—BURSTING, Calc. cau., Caus., Hell., Lyc., Merc., Mur. ac., Nit. ac., Pul., Sep.; with stitches, Lyc.

—CRAMP-LIKE, partly, and partly sticking, and feeling as if stopped by a swelling, MERC. SOL.

—DEEP-SEATED, Coch. (l.), Pho.; evening while walking, Lach.; with pain in forehead, Sang. (l.).

—DRAWING (including PULLING), Bell., Merc.; WITH hard hearing, Lach.; heat, Sep.; feeling as if moisture came out of ear, Mil. (l.); EXTENDING FROM front side of head, All. c. (l.); front of neck, Jac.; within outward, Sep., Sul. ac.; throat, All. c. (r.); in INNER PASSAGES, Cyc. (r.).

—DRAWN, as if, Verb.

—DULL, Pau. p. (l.); causing dulness, Merc. i. f.

—EXPANDING (including STRETCHING), Calc. cau., Con., Kre., Nit. ac., Rhodo., Spig., Til., Vio. o.

—EXTENDING from one EAR to other, on rising in bed, Ment. pi.; downward, Thu.; not pressing together, but going lengthwise, Mil. (l.); to FOREHEAD, Pte. (r.); lower parietal, while sitting after dinner, Indg.; to temple, For.; after dinner, while sitting, Indg.; to temples and top of head, Lac. ac.; to left ORBIT, Hur. (l.); out from, K. iod.; to CHEEK, Stram. (l.); to jaw, Spigg. (l.); lower jaw, Com.; to TEETH, Itu.; to THROAT, Hæm.; afternoon, Fago. (r.); from NECK, Thu.; to lower part of neck, along course of carotid artery, < inclining head to right, Lepi.; along side of neck, Hæm. (r.); to SHOULDER, Rum. c. (l.).

—HAMMERING, Thu.

—INFLAMED, as if, Merc. sol. (l.).

PAIN, INTERMITTENT, Tarent.

—KNOCKING, Anac., Bar. c., CALC. OSTR., Mag. m., Mur. ac., Nat., NIT. AC., Pho., Rhu. t., Sil., Spig.

—LANCINATING, Aur. s., Cad. s., Bell., Cit. v., Cham., Crot. c., Der., Fer. io., Gam., Hur., K. iod., Lepi., Mar., Pb., Nx. v., Rap.; in MORNING, Tarent.; when WALKING or SPEAKING, Mang.; > FOOT-BATH, Tarent.; WITH or without matter, Bell., Cham., Chin.; with hard hearing, Con.; sudden temporary deafness, as if something had fallen before the ear, < blowing nose, Chin., Con.; ACUTE, on stooping, turning head, Ment. pi.; extending to teeth on that side, when writing, Ment. pi. (l.); DARTING, Chin., Nx. v.; EXTENDING from root of nose, Elap.; to shoulder, Lepi.; outward, Æth.

—NEURALGIC, Iod.; in evening, Cit. v. (l., then r.); during gastric symptoms, Tarent.

—PULSATING (including THROBBING), Anac., Angu., Bar. c., Bell., Berb., Cannab. i., Caus., Cinch., Coc. c, Con., Grap., Ig., K. ca., Kre., Lyc., Mag. m., Merc. i. f., Nit. ac., *Pho.*, Rhe., Rhodo., Rhu. t., Sep.; at night, Rhu. t.; as from an abscess, Gam.; suppurative, < boring in, Anac.

—PULLING, see DRAWING.

—RHEUMATIC, Merc. i. r. (l.).

—SHIFTING from ear to orbit, Hur. (l.).

—SPASMODIC, Cioc. (r.); as if it would be pressed asunder, Spig. (r.).

—STRETCHING, see EXPANDING.

—SUDDEN, when walking in open air in afternoon, Chin. s. (l.).

—THREADS, about length of finger, from deep in the head, like, remaining in single spots about size of a pea or hazel-nut, All. c.

—TWITCHING, Am. car., Am. m., Anac., *Angu.*, Bar. c., Bell., Calc. a., Calc. ostr., Cannab. i., Carb. v., Cina, Cle., Con., Dig., Dro., Fl. ac., Hep., Lyc., Mang., Merc., Mur. ac., Nat. m., Nicc., Nit. ac., Nx. v., Petrol., Pho., Pho. ac., Plat., PUL., Rhodo., Sars., Sil., Spig., Spo., Sul. ac., Tab., Val.; acute, Rhodo. (r.); extending from teeth, Pul.

—TUGGING, Chin.

—ULCERATIVE, Anac., Cap., Caus., Cic. v., Fer., Grap., K. ca., Mag. c., Mang., *Mur. ac.*, Pso. (l.), Sars., Sep., Spo.; on BITING teeth together, *Anac.*; DEEP-SEATED, in morning, Sars. (l.); < boring into ear, Mur. ac. (l.).

—WOUNDED, as if, Anac., Caus., Lyc., Mag. m., Mar., Sep., Spo.

PARAPHLEGIA. Merc. (l.).

PIERCING. Calc. c., Canch., Glo. (r.); <cough, Nit. ac.; outward, K. ca.

PIMPLES, with itching. Berb.

PINCHING. Agn., Am. car., Anac., Ars., Asar., Bell., Bry., Carb. a. (l.), Carb. v., Caus., Colch., Coloc., Con., Crot. t., Der., Dro., Dul., Euphm., Gran., Guai., Iod., K. ca., Kre., Lau., Lyc., Mang., Mar., Meny., Merc., Mez., *Mur. ac.*, Na. m., Nit. ac., Nitr., Nx. m., Nx. v., Pho., Plat., Ran. s., Rhe., Rhodo., Rhu. t., Saba., Sabi., Sep., Spig., Stan., Stap. (l.), Sul., Thu. (r.); in MORNING, Na. c. (r.); AFTERNOON, Aran. d. (r., then l.); NIGHT, *Bry.*; after HICCOUGH, Bell. (r., then l.); WITH aching, Nx. v.; cracking, Na. c.; stitches,

Sam.; stitches behind the ear with decrease of the pain, Nitr. (l.); drawing pain in articulation of jaw, Na. c.; DEEP-SEATED, Fer. ma., Mur. ac. (r.), Sabi.; jerking pinching, Mur. ac. (l.); EXTENDING inward, Dul.; outward, Carb. v., Rhodo.; itching, Rhe.; PRESSING OUT, Caus.

PISTON working up and down, sensation of. Aml. n.

PLUGGED SENSATION. See STOPPED SENSATION.

PRESSURE. Acon., Aconin., Anac., Arn. (l., then r.), Asaf. (l.), *Bell.*, Berb., Calc. c., Calc. ostr., Calc. ph., Cannab., Carb. v., Carl., Coc. c., Crot. t., Eupi. (r.), Fl. ac. (r.), Glo., Hell., Indg., Ip., Kis. (Lach.), Lau., Lyc., Merc. i. r., Mur. ac., Na. c., Na. slfc. (r.), Nit. ac., Nx. m., Nx. v., Olnd., Petrol., Pho., Rhe., Rut., Saba., Sars., Sep., Sil., Spig., Spo., Tarax., Tel., Thu., Verat.; in MORNING, Verat. (r.); when waking, Sep. (l.); 6 A.M., after getting out of bed, Rum. c. (r.); EVENING, Hype. (r.), K. bi.; when CHEWING, Seneg. (r.); from PUSTULE, Gas. (l.); in warm ROOM, Pho.; (AGG.), 9 A.M., Na. slfc.; cough, Nx. v.; motion of lower jaw, Na. m. (l.); (AMEL.), in cold, Pho.; WITH heat in head, Jat.; fever-stage, *Asar.*; ACUTE, Rut.; on going into open air, Mang. (r.); ASUNDER, Cannab., Con., Grap., Hell., Nit. ac., Par., Pul., Rhodo. (l.), Spig., Spo.; BUBBLING PRESSING ASUNDER, K. n.; DEEP-SEATED, Mang. o. (l.); on which he was lying, Bar. c. (r., then l.); > stirring finger in, Fl. ac. (l.); DULL, Chel.; as from a FINGER, Acon. (l.), Rhe.; HOT, < inserting finger, Rut.; INTERNAL, Tarax. (l.); with tension, Calc. cau.; as if every thing would be pressed together, almost cramp-like, Dro. (r.); INWARD, Sep.; alternating with a sensation as if torn out, alternately with orbits, *Bell.*; from OPPOSITE SIDES, K. n.; OUTWARD, Berb., Caus., Chel. (r.), Guare., Ir. v., K. n., Merc., Mur. ac., Par., Pru. s., Pul., Sep.; from 10.30 A.M., Hydrs.; while straining at stool, Sep.; with cough, Nx. v.; rhythmical, Mur. ac. (r.); as if something must be torn from within, Lil. t.; followed by tickling, Chel. (r.); as from a PEG or PLUG, Anac., Cannab. s., Nx. m., Olnd., Par., Spig., *Spig.* (l.); as from a dull POINT, Hydrphb; SCRAPING, Rut.; STITCHING, evening, Berb. (r.); SUDDEN, acute, Con.; TENSIVE, Dig. (l.); as if something were pressing THROUGH, Calc. ostr.; TICKLING, < inserting finger, Rut.

PRICKING. See STITCHES.

PRICKLING. Dul.; itching, Spig. (r.).

PUFFING, from pulsation of temporal arteries. Benz. ac.

PULLING. See DRAWING.

PULSATION (INCLUDING BEATING and THROBBING). Aloe, Alum., *Aml. n.*, Am. m., Bell., Bell. (r.), Berb. (l.), Bon., Bro., Cact., Cala. (r.), *Calc. c.*, Calc. ostr., Calc. ph., CANNAB. I., Cannab. s., Carb. ac., Carbn. o. (l.), Carb. v., Caus., Chel., Chin., Coloc. (< l.), Con., Dig., Gam. (l.), Glo. (r.), Grap., Hydrc. (r.), (Ig.), K. ca., K. n., Lyc., Mag. m. (r.), *Merc. c.* (< l.), Mur. ac., Na. c. (l.), Na. m., Ol. an., Op., Pho., Plan. (l.), Pte. (r.), Rhe., Rhodo. (l.), Sel. (r.), Sep., Sil. (r.), Spig. (l.), Sul., Tel., Zn.; FORENOON, Coca; EVENING, Cob. (l.), Ind., Physo.; on falling asleep, Sil.; in bed, Thu.; NIGHT, Dig. (< l.), Am. m. (r.); when lying on ear, Am. car. (l.), Bar. c., K. ca. (r.), Sil.; in OPEN AIR, Acon. (l.); ASCENDING steps, Gas.; after BREAKFAST, Zn.; during internal COLDNESS, Amyg.; after DINNER, Carb. a.; on LYING on left side, Bar. c. (l.); when SITTING, Am. m. (r.); on STOOPING, Rhe.;

after WALKING, Pho.; while WRITING, Rhe., Zn.; (AGG.) by exposure to air, Pte. (r.); eating, Grap.; (AMEL.), by pressure, Carb. a.; stooping, Grap.; turning over, Am. car. (l.); WITH cold sensation in ear, Berb.; creaking when lying on affected ear, and with pimples and pustules in external ear, Spo.; discharge, Tel.; dryness, Berb.; hard hearing, Hep.; otalgia, at night, Rhu. t.; rushing, Coloc.; ACUTE, deep, extending outward, Merc. i. f. (l.); ALTERNATING with singing, Caus.; after ringing and tearing, Plat.; as if they would BURST outward, Cact.; DEEP-SEATED, when lying on ear, Bar. c. (r., then l.); EXTENDING inward, Berb.; HEAVY, Rum. c.; PREVENTING sleep, Cact. (r.); RHYTHMICAL, lying on, Coc. c.; SHAKING EYES, Sil.; SLOWER than pulse, morning after waking, Grap.; SYNCHRONOUS with pulse, Coc. c. (l.), Rum. c.; WAVE-LIKE, > holding hand over eyes, Spig.; TWO pulsations, then heat rushing out, Ol. an.

RAILROAD-TRAIN going through brain, sensation of, > alternate days, Chin. s.

RAWNESS. 2.30 P.M., Ol. an.

ROARING would occur, sensation as if. *Mez.*

ROLLING back and forth, it seems as if something were, on shaking head. Rut.

ROUGHNESS. Acon. l., Rhu. v.

SCRAPING. Lyc.; as from an ear of corn, Pb.; as from a feather, Ol. an.; as from the turning of a blunt stick, Rut.

SCRATCHING. Pb. (r.), Pho. ac., Rut.; like a bristle, Pb.

SCREWING, extending to shoulder, morning on waking, Daph.; twisting, evening, Nx. v.

SENSATION in inner passages lost. Mur. ac.

SENSITIVENESS, and of adjoining parts, with otitis, Merc.; to draught, wind, etc., Cham., Lach.; to touch, with discharge, Carb. v., Cham.; to touch, with otitis, Mag. c.

SHOCKS. Anac., Bell., Calc. ph. (l.), Con., Na. m., Na. m. (l.), Nx. v., Spig.; WITH contractions, Bell.; FOLLOWED by ringing, Na. m.; as of distant ARTILLERY, Bad.; ELECTRIC, Cannab. i.

SHOOTING, Apoc. c, Arum d., For. (l.), Iod. (r.), Phyt., Plan., Sep. (r.), Sil. (l.), Stram., Trom. (r.); AFTERNOON, Na. ar. (r.), Trom.; 3 P.M., Physo. (l.), Trom. (r.); 3.30 P.M., Trom. (r.); 6.45 P.M., Physo.; 10 P.M., For. (l.); after BREAKFAST, Arum d. (r.); when LYING ON RIGHT SIDE, Pte. (r.); when SNEEZING, Calc. c.; < DESCENDING stairs, Iod. (r.); WITH aching, Lob. i. (r.); deafness, Bell.; as if ABSCESSES were forming, < walking, Ment. pi. (< l.); ACUTE, making him cry out, Nx. v.: DEEP-SEATED, Fer. ma.; EXTENDING to chin, Bell.; inward, Æth., Alum., Am. m., Arn., Berb., Carb. v., Dro., Dul., Grap., Hyos., K. bi., K. iod., Mag. c., Meny., Na. c., Na. s., Nx. v., Pho., Rhu. t., Stro., Tarax., Ton.; outward, Æth., Alum., Am. m., Ars., Cannab. s., Dul., Glo., Gran., K. ca., Lau., Nicc., Rhodo., Sep., Spo., Stro., Tarax., Vio. o.; to spine, Pte. (l.); to temporal region, Eupi. (l.); upward, Lil. s.; OBTUSE, Cact., Mang., Meny., Na. m., Nitr., Plat., Sars.; STICKING, Æth. (l.); SUDDEN, Pho.; TEARING, Tep.

SMARTING. Bor., Cannab. s., Caus., Cic. v., Lach., Lyc., Ol. an., Sep., Spo.; corroding, Arg., Dro., Plat.; itching, Lyc.

SNAPPING. Lac. ac., Tarent. (r.); MORNING, on rising, when chewing, Aq. p. (r.); on OPENING MOUTH, Dul. (l.); as of ELECTRIC SPARK, Ambra.

SOMETHING BEFORE, sensation of. Acon., Alum., Angu., Bry. (l.), Cala., Card. b., *Chin.*, Cocc. (r.), Cyc. (r.), K. iod., Merl. (l.), *Pho.*, *Sul.* (l.); during MENSES, > boring with finger, Mag. m.; on blowing NOSE, Con.; INTERMITTENT, Pho. (r.).

SORENESS. Bry. (r.), Calc. ph., Jug. r., Mag. s. (l.), Mos., Pte. (r.), Sep., Sul. (r.); EVENING, after removing earwax, when walking, Bor. (l.); on BORING in with finger, Bor.; after itching, Arg.; on TOUCH, Act., Ery. a. (l.), Mag. c. (l.), Mag. v. (r.), Merc. i. f., Spo. (r.), Zn. a. (l.); as if BEATEN, ARN., Cham., *Cic. v.*, *Cina*, Crot. h., RUT.; BURNING, Pho.; EXTENDING from throat, Lith.; down neck, Bap.

SPASMODIC sensation in and around the ears. Ran. b.

SQUEEZING. ARN., Bell., Carb. a., Chel., Dro., Grap., K. ca., *Rut.*, Sars., Spig., Thu., Zn.; sensation as if something were squeezed out, Thu.

STARTING. Benz. ac.; jerking, with otitis, Nitr.

STITCHES (including PRICKING and STINGING. Compare with TEARING). Æth. (r.), Aga. m. (l.), Aloe (l., then r.), Alum., Am. car. (r.), Aur. m. n., Bar. c. (l.), Bart., Bell., Berb. (r.), Bor. (l.), Bro. (r.), Bry., Cala., Camph, Calc. c. (l.), Calc. ostr., Carb. a., Cham., Chel., Coc. c. (l.), Colch. (l.), Cob. (l.), Con., Cup., Dro. (r.), Dul., Euphr., Fer. p. (r.), Fl. ac., For., Gam., Glo. (r.), Hip. (l.), Hur., Ig., Iod., Indg., Jat., Jat. (l.), Kalm., K. bi., K. bi. (l.), K. ca., K. n., Kin. (r.), Kre., Lach., Lact., Lepi., Lepi. (r.), Lau., Lip., Lyc., Mag. m. (r.), Mag. r. (l.), Meny. (r., then l.), Merc. c. (l.), Merc. i. f. (l.), Mil. (l.), Mur. ac., *Na. c.*, Na. m., Na. slfc. (r.), Nicc. (l.), Nit. d. s., Nit. ac., Nit. ac. (r.), Nx. m., Ol. an., Pb., Petrol., Phe., Pho. ac., Pho., Plat., Ple., Pso. (l.), *Pul.* (l.), Ran. b. (l.), Rap. (r.), Rhu. v., Rut. (r.), Saba. (l.), Sam. (r.), Sarr., Sep., Sep. (l.), Sil., Spig., Stap., SUL. (l.), Tab., Tarent. (r.), Thu., Til., Verat. (l.), Vesp. (r.), Wild. (r.), Zn. (r.); MORNING, All. c. (r.), Ars., Fer. (r.), For., Nx. v.; when washing in cold water, Bor.; FORENOON, Chin. s. (l.), K. bi. (r.), Mag. c. (l.), Na. m., Nx. m., Ple., Sarr. (r.); 10 A.M., Ir. f., Mag. s.; NOON, Ammc. (r.), Chin. s. (r.), Pso.; AFTERNOON, Bry. (l.), Carbn. s. (r.), For. (l.), Merc. c. (l.), Ple.; EVENING, Aq. m. (r.), Berb., Bor. (l.), Chin. s. (r.), Grap. (l.), Hype. (r.), Junc., K. ca., Merc., Ox. ac., Ran. b. (l.), Stap. (r.), Tarent.; in bed, K. iod., Thu.; after eating, Grap. (l.); 1 P.M., Grap. (l.); 4 P.M., Kalm., Na. c. (r.); 5 P.M., Berb. (r.); 8 P.M., Na. slfc. (r.); 8 to 10 P.M., Phel.; 9 P.M., Carbn. s. (l.); NIGHT, Alum. (r.), Cop. (l.), Kalm. (r.), Pho.; when awaking, Carbn. s. (l.); during toothache, Hell. (r.); when going to BED, Fer. p. (r.); during stroke of BELL, Pho. ac., Mag. m. (l.); after CRAMP in ears, Thu.; after DINNER, Carbn. s. (r.), Zn.; from DRAUGHT OF AIR, Camph.; while EATING, Verb. (l.); when pressing FOREHEAD, Nit. ac.; when moving JAW, Pho. ac.; during her own SINGING, Pho. ac.; while STANDING, Mag. s. (l.); when STOOPING, Merc. sol.; when TURNING head, Chin. S. (r.); when WALKING, Ammc. (r.), Bor. (l.); (AGG.), evening, Na. m.; open air, Tab.; stooping, Lepi. (r.); (AMEL.), putting finger into ear, Pho. ac.; rubbing, Ol. an.

—WITH stitches around EAR, Con., Vio.; stitches behind, Am. car., Bell., Kalm. (r.); boring inward, Mag. c.; feeling as if closed, Glo.; hard hearing, Am. m.; inflammation, Alum. (l.), Merc., Merc. c.; sensation as if too narrow, Lyc.; pain as if it would burst, Lyc.; pinching, Sam.;

roaring, Caus., Nit. ac.; musical sounds, Pho. ac.; swelling, Kre.; tearing, Cham., Pb.; tearing in ear, catarrh of Eustachian tube, redness and swelling of meatus, Pul.; boring tearing, Hip. (l.); EXCITEMENT and delirium, Stram.; weeping and weakness, Sil.; loud lamenting, Sep.; crying out, Nx. v.; HEADACHE, < at night, Cyc.; headache, and stitches in parotid gland, K. bi. (l.); pain in FACE, Pho. ac.; pain in jugular (zygomatic bone), Bro.; DIPHTHERIA, K. bi.; COUGH, Nit. ac.; with FEVER stage, Calc., *Grap.*, Pul.; beginning of chilliness, Gam.; sudden general sweat, Caus.

STITCHES, FOLLOWED by hard hearing, Cap.; straining in ears, Lact. (l.); pain in arms, Kalm.

—ACUTE, Cocc. (l.), Na. c.; piercing deep, Bry.; making him start, Mag. m., Ton.

—ALTERNATELY in each, on entering open air or house, Bry.

—AWL, as with an, Mag. m. (l.).

—BORING, Caus. (r.).

—BURNING, itching, Tarax. (r.).

—BURROWING, Na. m.

—DEEP-SEATED, Jat. (l.).

—DIGGING, Berb. (r.).

—DRAWING, Berb. (r.).

—DULL, Nicc., Thu. (r.).

—EXTENDING FROM PALATE, Cob. (l.); TO DRUM, Dul. (l.); to EYE, with a sensation of a wind streaming into ears, Pul.; to HEAD, Bry.; INWARD, Am. m., Arg., Arn., Carb. v., Dro. (l.), Nx. v., < on coming from cold air into warmth, not the reverse, Na. slfc. (r.); in and out, K. ca.; to LOBULE, Pho.; to NOSE, Sil.; OUTWARD, Alum., *Am. m.*, Calc. c., Cannab. s., Con., K. ca., Lau., Mang., *Na. c.*, Sil., Stro., Tarax., Vio. o., Zn.; to PARIETAL BONE, Ran. b. (r.); to TEMPLE, Indf.; to HEAD, with red, rough tetter in front of ear, and oozing and bad smell behind ear, Olnd. (l.).

—FINE, Berb. (l.), Colch., Ox. ac. (l.), Pb. (r.); with itching, Berb.

—HOT were streaming from, as if something, Æth. (< r.).

—INTERMITTING, Benz. ac. (r.).

—LARGE, shooting, with fretfulness and vexation about trifles, Cham.

—PRESSING, Fl. ac. (r.).

—PULSATING outward, Glo. (r.).

—NAIL, as from a, while sitting, Berb. (r.).

—RHEUMATIC, Lyc. (r.).

—SLOW, Sang. (l.); broad, extending inward, Dro. (l.).

—SUDDEN, extending into sternum and left side of neck, Cocc.

—TEARING, Lyc., Nx. v.; with external swelling, especially in children, Zn.

—TWINGING, Colch. (l.).

—TWITCHING, Aga. m. (r.).

STOPPED sensation (including OBSTRUCTION and PLUGGED sensation). Acon. (l.), Æth. (<l.), Aga m. (l.), Anac. (l.), Angu., Arg. (r.), *Arg. n.* (l.), Asar., Ath., Aur. m. (l.), Berb. (l.), Bor., Bry., *Carb. v.*, Caus. (r.), Cham., *Chel.*, Chin. s., Chlf., Cinnb., Coc. c. (l.), Cod., Colch. (r.), *Con.*, Conin.,

Crot. h. (r.), Cyc. (r.), Dig., Dios., Gas., *Glo.*, Grap., Guarc., Hur., Hydrc. (l.), Hydrs. (l.), Indg., K. bi. (l.), K. ca., Led., Lepi., Lyc., Merc. c., MERC. SOL., *Mez.*, Mil., Na. c. (l.), Na. m., Nit. ac., Ol. an., Op., Pho., Physo., Ple., PUL., Rap., Rhu. t. (r.), Run. c. (l.), Saba., Sec. c., Sel., Seneg., Sep., Sil., Spig. (l.), Sul., Tab., Tel. (r.), Tep., Thu., Til. (r.), Upas, *Verb.* (l., then r.); MORNING, Ant. cr., Caus., Thu. (r.); on rising, Sil.; FORENOON, Tel.; AFTERNOON, Mil., Na. m.; EVENING, Ant. cr. (r.), Ham. (r.), Thu. (r.); in bed, Sel.; 3 P.M., Jac.; 8 P.M., Dios.; 10 P.M., Tel. (< l.); after DINNER, Mil.; after LYING on it, Sel. (l.); on READING aloud, Verb.; after STITCHES in glands, Berb.; while TALKING, Meny.; on WALKING, Colch.; (AGG.), excitement, Dig.; lying on ear, Coc. c. (l.); (AMEL.), after breakfast, Ant. cr.; afternoon, yawning, Na. m.; inserting finger, Sel. (l.); WITH drawing, Lach.; good hearing, Mez.; heaviness in and before ears, Carb. v.; roaring, Grap., Hell., Merc. c., Seneg., Sep.; BEFORE ears, Coloc. (l.); DEEP-SEATED, in evening, Lim. (r.); as from GREASE, and again opening, Bov.; HISSING THROBBING, Coc. c., Hep., *K. ca.*, Nx. v., Ol. an., Petrol., Pul., Saba., Sul., Thu.

STRAINING after stitches, Lact. (l.); with tearing in alternating jerks, Mez.

STUFFED sensation, Æth. (r.), Cannab. i. (r.), Carbn. s., Cot. (r.), Nico., Sul.; forenoon, Pso.

SWARMING (of animals), sensation < lying still, Ant. cr.

SWELLING. Ant. cr. (l.), Apis., Ars., Calc. c. (r.), Calc. ph., Cannab. s. (r.); Caus., Chlol., Cist., Crot. c. (r.), Ery. a. (l.), Glo. (r.), Grap. (l.), K. ca., Pho. ac., Jug. (r.), Med., Nit. ac., Pic. ac., Pte. (r.), *Rhu. t.* (l.), Rhu. v., *Tel.* (l.), Ust. u., Zn. ac. (l.); WITH discharge, Caus., Cist.; itching and discharge, Tel.; inflammation, Kre., Nitr.; stitches, Kre.; EXTENDING half way up cheek, Cist.; over parotid gland to Zygoma, Bry.; HOT, Bor.; INFLAMMATORY, *Bell.*, Bor.

SUPPURATION. Caus., Hep., Lyc., Merc., Sec. c.; with itching and hard hearing, Am. car., Bov.; sensation of, on touch, Coc. c. (l.).

TEARING (including LACERATION. Compare with STITCHES). Acon. (l.), Æth. (r.), Aga. m. (r.), Ambra (r.), Am. m. (r.), Arg. n. (r.), Arn., Arum t., Bov. (r.), Cad. s., Calc. c. (l.), Calc. ph., Camph. (l.), Cannab. i. (r.), Canth. (r.), Carb. a. (l.), Carb. v. (r.), Caus. (l.), Cham., Chel. (l.), Con. (r.), Cup. (r.), Cyc., Guai. (l.), K. bi, *K. ca.*, K. iod., Lach., Lachn., Lau., Lyc. (r.), Mag. m. (r.), Mep., Merc., MERC. SOL. (r.), (l.), Merl. (r.), (l.), Mur. ac. (r.), (l.). Na. c., Nicc. (r.), Ol. an., Par., Pb. (r.), (l.), Petrol., Pho., Plat., Rat., Rat. (r.), Rhodo. (r.), Sabi. (l.), Sil., (Squ.), Stram. (r.), Stro. (r.), Sul. (l.), Tab. (r.), Tarent. (r.), Til., Ton. (l.), Verb. (r.), *Zn.*; MORNING, Ant. t. (r.), Mang. (r.), Sars. (r.); in bed, Carb. v.; FORENOON, Mag. c. (l); while sitting, Pho.; 9 A.M., Elap. (l.); AFTERNOON, Bov. (l.); about NOON, Sul. (l.); EVENING, Alum., Indg., Ton. (r.); in bed, Thu.; 7 P.M., Zing. (r.); after DINNER, Carb. a., Phe.; INCLINING body to right, Mag. m. (l.); (AGG.), cough, Pul.; (AMEL.), pressure of hand, Alum., Carb. a.; WITH sensation as if a cool breeze blew against it, Stram.; crawling, Zn.; creeping, Sul. ac., Ton.; discharge, after measles, Colch.; inflammation, Merc.; itching, Rhe.; stitches, Cham., Con., Pb.; tearing in ear, catarrh of Eustachian tube, redness and swelling in meatus, Pul.; straining, in alternating jerks, Mez.; twitching, roaring and ringing, from one's own singing, Pul.; FOLLOWED by discharge, K. ca.; ALTERNATELY in each, Aph., Chen.; BORING, with stitches, Hip. (l.); DARTING, Pul.; DEEP-SEATED, Gam., Mez. (l.), Pb. (r.), Sul. ac. (l.), Ton. (l.); evening, Merl. (l.); during menses, Merc. sol. (l.); on rising from stooping, Ant. t. (r.), Mang. (r.); extending into parietal bone, forenoon, Indg.; DRAWING, Aph., Coc. c. (l.), Merl. (l.), Zing. (r.); EXTENDING from face, Mil.; backward, Bell.; inward, K. ca., Lau., Nx. v., Ton., Verb. (l.); outward, Bar. c. (l.), Dul., Til. (r.); into external cartilage, K. ca.; to cheek, Anac. (l.); into head, *Sul.* (l.); to jaw, Merl. (l.); through occiput, Ambra; to teeth, Ol. an.; in afternoon, *Chel.* (r.); to temples, Eupi. (r.), Lach.; FINE, Grat. (l.); intermitting, 8 A.M., while standing, Na. c. (r.); internally, Cyc. (l.); INTERMITTING, 8 A.M., while sitting, Na. c. (r.); ITCHING, Caus.; RHEUMATIC as if broken, Mez. (l.); STICKING, K. bi., Mar. (l.); forenoon, while standing, Pb. (r.); afternoon, Æth. (r.); as if in bone, Rap. (l.); extending outward, Gran. (l.); THRUSTING, Spig. (r.); TWITCHING, Mag. m. (l.), Mez. (l.), Pul., Tab. (r.); as if being TORN OUT, Bell., Cannab. i., Bry. a., Par., Sap. (l.); and at the same time in bone above and beneath patella, K. ca.; alternating with orbits, as if alternately torn out and pressed in, *Bell.*

TENSION. Alum., Am. car., ASAR., Aur., Bov., Calc. cau., Calc. cau. (l.), Carl., Coc. c., Colch., Con., Dig., Dro., Euph. p., Euphr. (l.), Glo., Kre., Lact., Lachn., Lyc., Ment. pi. (< r.), Mez., Nitr., Nx. v., Pb. (r.), Spo., Stap., Thu., Verat., Vio. o.; MORNING and AFTERNOON, Hydrs. (r.); when BRUSHING HAIR on occiput, Ars. s. f. (r.); WITH want of ease, Caus.; CHANGING from temples, waking from sleep, < morning, Glo.; INTERNALLY, with pressure, Calc. cau.

THREAD drawn through, sensation of. Rhu. t.

THROBBING. See PULSATION.

TICKLING. Acon. (r.), Aga. m. (r.), Ambra, Am. m. (r.), Ars., Bor. (l.), Bro., Canth. (l.), Carbn. s, Chel. (r.), Chin., *K. ca.*, K. iod., K. n., Lach., Mag. m. (r.), Mang., Mur. ac. (l.), Na. c., Petrol., Saba., Sang., Sul., Sul. ac., Tab., Til. (r.), Ton. (l.), Zn. (l.); EVENING, Cala. (r.), Ple.; 6.30 P.M., Mag. c. (r.); 10 P.M., from sound of voices, Rum. c.; after DINNER, Phe.; after painful PRESSING OUT, Chel. (r.); (AMEL.) boring in with finger, Mag. m.; after eating, Acon.; WITH pleasant warmth, Mag. m.; FOLLOWED by stiffness of jaw, Petrol.; CRAWLING, after dinner, Lau.; with ringing, Cinch.; as from EARWAX, Crot. h.; as from a VEIL over, Pho.

TINGLING. Anac. (r.), *Bell.*, Brach., Cent., Chin. s., Fer. ma., Lachn., Lol. Osm. (r.), Saln. (r.), Sep., *Sul.*, Sul. i., Thu. (l.); noon, Stry.; > BORING in with finger, Lachn.; WITH pain, Osm.; TREMULOUS, Pul.

THRUSTS, ACUTE. Nx. v.; forenoon, Gen. (l.); DULL, back and forth, Nx. m.; coming from both sides, as if two plugs were penetrating to meet in centre, Anac.

THUMPING, afternoon on entering room from open air. Thu.; < lying on either side, Na. h.

TREMBLING on rising. K. ca.; after hearing bad news, Saba.; as from rush of blood to head, Petrol.

TWINGING. Aloe, Anag. (r.), Arg. n., *Bar. c.*, Carb. v. (l.), Caus., Coloc. (r.),

Crot. t. (l.), Dul. (r.), K. n., Merc. sol., Mez. (< r.), Par., Pru. s. (l.), Stap. (l.); AFTERNOON, Aran. d. (r., then l.); EVENING, Aloe, Carb. v. (r.); ACUTE, morning before rising, Fer. (r.); DRAWING, Coc. c. (l.); EXTENDING OUTWARD, Carb. v.; SPASMODIC, deep in, *Crot. t.* (l.).

TWITCHING (including JERKING). Æth., Am. car. (l.), Bar. ac. (l.), Bar. c. (l.), Bov. (l.), Calc. c. (r.), Calc. p., Cannab. i., Cast. eq. (l.), Hep., K. ca., Mag. m. (r.), Manc., Op., Petrol., Pho., *Pul*, Sil. (l.), Sul. ac. (r.), Thu. (r.); MORNING, Ant. t. (r.), Mang. (r.); on waking, Nx. v.; EVENING, Nx. v.; on RISING, K. ca.; WITH cold sensation, Plat.; external heat, Pul.; tearing, roaring and ringing, from one's own singing, Pul.; DRAWING, Cle.; EXTENDING inward, Pho.; to eye and lower jaw, Spig.; to lower jaw, Nit. ac. (r.), Caus.; as if SOMETHING WERE TWITCHED OUT WITH A HOOK, 6 A.M., Na. m. (r.); PINCHING, Zn.; QUIVERING, extending to left corner of mouth, Thu.; STICKING, Nicc.; boring dull sticking, extending into throat, Spig.; SUDDEN, Chr. ox. (l.); TEARING, *Pul.*; THUNDERING, like distant cannonading, Plat.

TWISTING. Am. m. (l.).

UNPLEASANT SENSATION. Feru., < evening, Physo.; indescribable, Mos. (r.); from noise, Merc.; with crackling, Mos.

VALVE, sensation as of a, opening and closing at each step. Grap. (r.); leather-covered metal valve in motion, Aga. m.

WARM FEELING. Alum., K. ca., Mang., Meny., Mur. ac., Plat., Seneg., Sul. ac.; WITH fluttering, Mang.; itching, Berb.; redness, Plat.; feeling as if a thin skin were over it, Asar.; PLEASANT, with tickling, Mag. m.

WATER RUSHING in, sensation of. Sul.

WAVING, as from rush of blood to head. Petrol.

WHIRLING. Lyc., Merc. (l.), Nx. v.; evening in bed, Lact.; whirring, Calc. a., Meny., Pul.; quavering, Nx. v.

WIND in, sensation of. Bell., *Chel.*, Eupi., Meny., Mez., Mos. (l.), Pul., Rhu. t., Spig., Stan. (l.), Stap., Stram., Vinc.; WITH tendency to bore in, Mez.; deafness, Cocc. (r.); hard hearing, except for speech, Ign.; hissing, Dig.; pulsation, Coloc.; sounds penetrating the whole body, Ther.; sound of trickling water, Thu.; COOL, Stan., with tearing, Stram.

WORMS in. Rut.; sensation of, Guare., Pic. ac., Rhodo.

EXTERNAL EAR.

External Ear. ASLEEP SENSATION, Sul.

BLISTERS. See VESICLES.

BLUENESS. Nit. ox., Sant.

BROWN SPOTS. Cop.

BURNING. *Cle.*, Jug. r., Kre., Pic. ac., Upa.; 1 A.M., Mit. (l.); EVENING, Ars., Trom. (< r.); after COFFEE, Sul. (r.); after SCRATCHING, Ol. an.; WITH heat and redness, as if they had been frozen, Aga. m.; stitches on pressure and swelling, Nitr., Pho.; picking stitches in middle ear, Cle.; swelling and heat, Na. m.

CHILBLAINS, affections as from, caused by cold. Bell.

COLDNESS. Berb. (l.), *Calc. ostr.*, Dul., *Ip.*, Ir. fœ., *K. ca.*, *Lach.*, Mang., Meny., *Merc.*, *Plat.*, Seneg., Stap., Verat.; feeling of, Nit. ac.; COLD FEELING IN, Calc. ostr., *Ip.*, Meny., Stap., *Veratn.;* after HEAT, Berb., Lach., Merc.; in a WARM ROOM, K. ca.; WITH paleness, Verat. v.; cough, Verat.; hot stage, *Ip.;* ALTERNATING with heat, Verat.; as if WATER got in it, Meny.; as if cold water were running out and through ear, Merc.; like a cold WIND, Caus., Mang., Plat., Stan., Stap.; CATCHES COLD easily, Cham., Lach.

COMPRESSION. Thu.

CRACKS. Mar.

CRAMP. Ars., Juni., Thu. (r.); and in middle ear, evening, Ran. b. (l.).

CRAWLING. *Ant. cr.*, Bro., Sul. ac.; with swelling, Spo.

DENUDED. *Merc. sol.* (< r.).

DESQUAMATION. Bry., Cop., Rhu. t.

DRAWING. Con., *Olnd.*, Pho. ac., Stan. (r.), *Tarax.*, Vio. o. (l.); sensation as if drawn out of head, Cannab. s.; drawing-together pains and swelling, Caus.

DRYNESS, with hard hearing. Petrol.

ELEVATIONS. Sep.; like wasp-stings, brownish yellow, Cop.

ERUPTIONS. Aga. m., *Bar. c.*, Cic. v., *Petros.*, Sep.; WITH eruptions behind ear, Cinch., Cic. v., K. iod., Petrol., *Pho.*, Pul., Sep., Sil., Spo., Sul.; swellings, Sep.; CONFLUENT, Cop.; HUMID, and behind ear, Calc. ostr.; ITCHING, Thu. (r.).

ERYSIPELAS, with itching, heat, redness, and blisters. Mep.

FORMICATION. Arun. (l.).

FURUNCLES. Pic. ac., Pul., Spo., Sul.

GNAWING. K. ca.

HEAT. Agn. (l.), Angu., Arn. (l.), Asar. (r.), Berb., Calc. p., Canna, Cap., Chin., Cle., (Cocc.), Cyc., Der., Grat., Hep., Kre., Meny., Mur. ac., Ped., Petrol., Pho. ac., Pul., Ran. b., Spira., Sul.; EVENING, Bry., Tab.; after SCRATCH-

ING, Ol. an. (r.); WITH heat of middle ear, Calc. p.; partial deafness, Mur. ac.; itching and redness, Hep.; violent itching, K. ca.; redness, *Alum.*, Ant. cr., Carb. v., Kre., Mag. c., Mep., Na. m., Nit., Peti. (l.), Pho., Pul., Sep., Tab., evenings, Alum. (r., then l.), every evening, Carb. v. (r.); swelling, Na. m., Pho. ac., Pul., Zn.; twitching in ear, Pul.; anxiety and rigor, Ars; headache, Zn.; cold stage, *Acon.*, *Alum.*, Ars., Bell, Dig., *Merc.*, *Pul.*, Rhu. t.; cold feet, Kre.; FOLLOWED by coldness, Berb., Lach., Merc.; ALTERNATING with coldness, Verat.; BURNING, Arn., Cle., Kre., Nitr.; and in middle ear, Casc., morning in bed, (Cocc.) (r.); redness, itching, and swelling, < night, Ail.; heat in back of head, Atro.; redness of cheeks, Sang.; IN, Acon., *Alum.*, Ars., *Asar.*, Bell., Bry., *Calc. c.*, Canth., Grap., *K. ca.*, Merc., *Na. m.*, *Pul.*, Sabi., Sep., Sil.; ON, Acon., Aga. m., Aloe, *Alum.*, Augu., Ant. cr., Apis, *Arn.*, *Ars.*, *Asar.*, Bro., Bry., *Calc. c.*, *Calc. ostr.*, Camph., Canth., Carb. v., Cinch., Cle., Grap., Hep., *Ig.*, *K. ca.*, Kre., Mag. c., *Merc.*, *Na. m.*, Nitr., *Olnd.*, PUL., Rhodo., Saba., *Sabi.*, *Sep.*, *Spig.*, Spo., Zn.; SPREADING from, Olnd., *Sep.*; as from a WIND, Physo.

HERPES, extending from temple over whole ear down to the chest, at times throwing off innumerable scales, and again showing painful rhagades, with humid yellow discharges, forming scurfs, fetid, humid, intolerable itching, especially in the evening till midnight. Pso. (l.); humid herpes, with swelling of cervical glands, and livid, gray complexion, Kre.; and around ears, with ulcers, Calc. p.

INFLAMMATION. Carb. v., Jug. r., Kre., MERC. SOL.; WITH redness and great soreness, Mag. c.; WITH INFLAMMATION OF MIDDLE EAR, Bell., Calc. ostr., Merc.; and discharge of pus, Bell. (r.); and redness, heat, and swelling, Pul.; stinging and tearing, Merc.; red, hot, swollen, burning, proceeding from a PIMPLE in concha, with stiffness and pain in left side of neck, shoulder and arm, Kre. (l.).

ITCHING. *Arg.*, *Aga. m.*, Benz. ac., Berb., Calc. ostr., Carb. v., Caus., Chel., Con., Fl. ac., Grap., Hep., Manc., Mep., Mez., Na. p., Paeo, Phel., Rap. (l.), *Sil.*, SPIG., Spira., *Sul.*; 1 P.M., > scratching, Ol. an.; 6 P.M., Grat. (l.); EVENING, Calc. p.; after DINNER, Phe.; after RIDING, Calc. p.; in ROOM, Calc. p.; > SCRATCHING, Chel.; WITH blisters, redness, and erysipelas, Mep.; heat and redness, Hep.; redness, Nit. ac.; stitching, in evening, and swelling, Calc. c., Pho. ac.; swelling, Nit. ac.; as if they had been FROZEN, with burning redness, Aga. m.; VIOLENT, with heat and redness, K. ca.; with heat, redness, and swelling, < night, Ail.; forcing her to scratch until they bleed, Arg.

KNOTS. Spo.

LIVID COLOR. Carbn. o., Op.

MOTION rapid and irregular. Acon. (r.).

NUMBNESS. Plat.; extending to lips, Plat. (r.).

PAIN. Fer. (l.), *Petrol.*, Rhodo. (r.), Sul. (l.); NIGHT, Pho.; when lying on it, Hep.; WITH swelling, Carb. a.; DRAGGING, Anac., Rum. c.; TENSIVE, with swelling, Spo.; ULCERATIVE, Fer., Mag. c.

PALENESS. Amyg., Lau., Rhu. t.; with coldness, Verat. v.

PIMPLES. Am. m., Berb., Cala. (l.), Calc. p., K. ca., Mag. c., Mang., Merc., Mur. ac., Pso. (r.), Sul. (r.); > evening, Petrol.; WITH itching, Berb.; pustules, and pulsation in ear, which creaks when lying on affected side, Spo.; INFLAMED, SORE, Cannab. s. (r.); OOZING, with swelling, Spo.

PINCHING. Angu., K. ca.; < rubbing, Mang. (l.).

PRESSURE. Anac.

PULLING their ears constantly, < evening, Pul. (< l.).

PULSATION. Merc. i. f. (l.); while sitting after dinner, Indg.

PURPLE COLOR. Apis, Sec. c.

REDNESS. *Acon.*, AGA. M., *Alum.*, Ant. cr. (l.), Asaf., Bell., Bry., Calc. p., *Camph.*, Canth., Carb. v., *Cinch.*, Der., Glo., Grap., Hep., *Ig.*, Ind., Jab., Jug. r., K. ca., K. n., Kre., Lyc., Manc., Merc., Nitr., Nit. ox., Peti., Petrol., Pho., Plat., *Pul.*, Sep., Spira., Spo.; AFTERNOON, Canc. f., Na. m. (r.); EVENING, *Alum.*, *Carb. v.* (l.), Elap., Oena., Rap., Rhu. v., Sep., Spire., Tab., Tarent., Trom. Vesp.; WITH blisters, itching, and erysipelas, Mep.; heat, *Alum.*, Ant. cr., Carb. v., Hip., Kre., Mag. c., Manc., Mep., Na. m., Nitr., Peti. (l.), Pho., Pul., Sep., Tab.; evenings, Alum. (r., then l.); every evening, Carb. v. (r.); heat and itching, Hep., K. ca., as if they had been frozen, Aga. m.; heat, itching, and swelling, < night, Ail.; itching, K. ca., Nit. ac.; inflammation, Kre., Mag. c., and great soreness, Mag. c.; suppuration, Nit. ac.; swelling, Apis (r. and l.), *Pul.*; warmth, Plat.; cold stage, Bell., Pul.; BLUISH, Tel., TELL. (l.); and looking as if infiltrated with water, with discharge, Tel.; ERYSIPELATOUS, Ars., Rhu. t.; LURID, Rhu. t.; DARK SCARLET, > sneezing, Chlol.

SCABS. Bell., Bor., *Bov.*, Elap., Grap., Iod., Lyc., *Pul.*, Sarr., Sil., Spig., Spo., Sul.; with swellings, Ars.; humid, suppurating, and behind ear, Lyc.

SCURFS. Cinnb. (r.), Cop. (l.), Sul. ac. (l.); and behind ear, Hep.; and humid scurfs behind ear, Pso.

SENSITIVENESS. Lach.; to pressure, remaining long, Mag. c.; to touch, Mur. ac., Rap.; to wind, draught, etc., Cham., Lach.

SKIN OVER, feeling of a thin, with warmth. Asar.

SORENESS. *Acon.*, Bry., K. ca., Mang., Merc., Mur. ac. (r.), Petrol.; extending towards temples, For.

STIFFNESS. Hydrphb.

STITCHES. *Chel.* (r.), Fer. ma., Ol. an. (l.), Sul. ac. (r.); on pressure, WITH burning and swelling, Nitr. Pho.; swelling, Spo., and itching, in evening, Calc. c., Pho. ac.; EXTENDING inward, night on entering house, Thu. (r.).

SUPPURATION. Spo.; with redness, Nit. ac.; burning, and around ear, Cic. v.

SWEAT. Pul.

SWELLING. Anac., *Ant. cr.*, *Calc. c.*, Caus., Crot. c. (r.), *Grap.*, *K. ca.*, Kre., *Lyc.*, Merc., Na. m., Nitr., Nit. ac., Pho. ac., PUL., *Rhu. t.*, Rhu. v. (l.), SEP., *Sil.*, Spo., Zn.; WITH blisters about ears, Ars.; burning and heat, Na. m.; burning and stitches on pressure, Nitr., Pho.; crawling, Spo.; discharge, Bor., Cist., Sil.; eruptions, Sep.; copious earwax, Calc. ostr., Na. m.; heat, Na. m., Pho. ac., Pul., Zn.; heat, redness, and itching, < night, Ail.; inflammation and tearing stitches, especially in children, Zn.; itching, Nit. ac.; itching and stitching, in evening, Calc. c., Pho. ac.; pain, Anac., Caus., Carb. a., and in head, Pso.; drawing-together pain, Caus.; tensive pain, Spo.; oozing pimple, Spo.; redness, Apis (r. and l.), *Pul.*; scabs, Ars.; stitches, Spo.; tension, Caus.; on and around, and on side of face, swelling like ERYSIPELAS, then running over the scalp, horribly painful to touch, Phyto.

SWOLLEN SENSATION. Lach.

TEARING. Anac. (l.), BELL., Con., Mur. ac. (r.), Pho. ac., Sul. ac. (r.); and in middle ear, Mag. m., Rat.; in middle ear and cartilage, K. ca.; straining, Bov.

TENSION. Thu., Vio. o. (l.); with swelling, Caus.

TETTER. Sep.; and around ear, extending to external meatus, Cist.

TICKLING. Colocn. (l.).

TINGLING. Stry.

TWINGING, and in middle ear, Asar.

TWITCHING. Anac., Bor., Plat., *Pul.*; and in middle ear, Dig.; cramp-like, *Cina.*

ULCERS. Bell.; with discharge, Merc.; and around ear, with herpes, Calc. p.; sensation of ulceration, Fer. (l.), K. ca. (r.).

VESICLES. Ars., Ars. (r.), Mep. (r.), Pte. (r.), Na. p.; WITH redness, itching, and erysipelas, Mep.; COALESCING, Ars. (r.); DISCHARGING water, Pte. (r.); GANGRENOUS, Ars.; LENTICULAR, Rhu. v. (r.); PURULENT, Ars. (r.); filled with SERUM, Rhu. v.; SURROUNDED BY INFLAMED BASE, Ars. (r.); TRANSPARENT, Alum. (r.); WHITE, Pte. (r.); on red base, Pte. (r.).

WHITE. See PALENESS.

ANTERIOR SURFACE. Redness, Bry.; near top, pimples, Coff.

ANTIHELIX. Painful pimple, Am. m. (r.).

ANTITRAGUS. ITCHING, BITING, Coc. c. (now r., now l.); DRAWING, *Spig.* (l.); PAIN, on pressure, extending into ear, Mur. ac.; PRESSURE, drawing into ear, Mur. ac.; PIMPLE, discharging like an ulcer, Spo. (r.); STIFFNESS, Kre. (l.); STITCHES on touch, Kre. (r.); biting, Coc. c. (l.); jerking, Kre. (r.); SWELLING, Kre. (r.); red, Spo. (r.); TEARING, Berb.; THICKENED, Bry.; TIP, sticking and tearing, Anac. (l.).

CARTILAGE. BRUISED PAIN, Rut.; DRAWING PAIN extending to neck, Pau. p. (l.); TEARING, and in external and middle ear, K. ca.; sticking near RIM, Elat. (l.).

CONCHA. ACHING as if it had been pressed against the head, Mos. (l.); extending into drum, then to occipital protuberance, Ib.; wipes BLOOD from, in morning, Calc. s. (l.); BRUISED sensation internally, *Arn.*; BURNING, K. bi., Merc. sol. (l.), Mur. ac., Na. m. (l.), Op., Pho., Spig. (l.), (r.), Wies.; COLDNESS, Thea; from a draught of air, Thea; of right, and burning heat of left, which extends beyond the temple, with pain apparently in tympanum with warmth, in evening, Na. n.; CONTRACTION, cramp-like, Anac. (l.); CUTTING extending to side of neck, Pau. p. (l.); DRAWING, Dro. (r.); cramp-like, Croc.; sensation of drawing towards back of, Asaf. (l.); ERUPTION, Chin.; HEAT, Arg. (l.), Lach., Na. m. (l.), Op., Pip. m. (l.); INFLAMMATION, Na m. (l.); erysipelatous, Tep. (l.); IRRITATION, Mos. (r.); ITCHING, Aga. m., Arg. (l.), Calc. c., Castor. (l.), Chel. (r.), Paeo., Ped. (r.), (l.), Rap., *Spig.* (r.), Sul. (r.), Wies.; evening when lying, Ped. (l.); burning, 9 P.M., > scratching, Phel. (l.); tickling, K. n. (l.); JUMPING, twitching, Aga. m. (l.); NEURALGIA extending to molars and cheek-bones, Thea; PAIN in evening, Mang.; PAPULES, Mur. ac.; PIMPLE with redness, heat, swelling and inflammation of external ear, and stiffness and pain in left side of neck, shoulder, and arm, Kre. (l.); PRESSURE, Bry. (r.), Coc. c. (l.), Cup. (r.), Lach.; drawing,

Sars. (r.); inward, Na. m. (l.); tearing, Sars. (r.); > pressure, Bism.; PULSATION, Fer. mu.; suppurative, < boring in, Anac.; REDNESS, Arn., Na. m. (l.); SCAB, Mur. ac.; SENSIBILITY LOST, Lach.; SHOOTING, Stan. (r.); SMARTING, Cannab. s.; SORENESS, Spo., Zn.; SQUEEZED sensation, Calc. p. (l.); STITCHES, Na. c. (l.), Rhu. v. (r.); evening in bed, Thu.; biting, > leaving warm bed, Coc. c.; fine, after dinner, Sul. (r.); squeezing, after dinner, while sitting, Thu. (l.); SWELLING, Arn., Na. m. (l.), Pho, *Sil.*, Tep. (l.); TEARING, *Cap.*, Chin., Cup. (l.), Hyos., K. ca. (r.), Lyc. (l.), Pho. ac. (l.), Thu. (l.); afternoon, Castor. (r.); evening, Indg. (r.), in bed, Thu.; 7.30 P.M., Mag. c. (l.); < pressure, Hyos.; pulsative, < boring in, Anac.; twitching, evening on lying down, > in bed, Ant. t. (r.); TICKLING, Sul. ac.; TURNED around, sensation as if, in morning, Mag. s. (r.); TWITCHING, Aga. m. (r.), Calc. a., Pho. ac. (l.), Spig. (r.), Upa.; ULCERATION, (Bry.); sensation of, in evening, Ant. t. (r.); VESICLES, Ars. (l.), Pho.

— ABOVE, deep in brain, pressure, with complete faint-heartedness, Aga. m.

— ANTERIOR wall, DRAWING, Pho. ac.; ERUPTION, Mos. (r.); ITCHING, Mos. (r.); > scratching, Chin. s. (l.); inflamed NODE with scab, painful to touch, Spo.; PRESSURE on motion, Pho. ac.; whitish SCALES fall from, in evening on scratching, Chin. s. (l.).

— BENEATH, acute gnawing, Dro.

— EXTERNAL, BORING, Lau. (r.); ITCHING, Am. m. (r.), Coc. c. (r.); PRESSURE, tearing, > pressure, Bism.; TEARING at 10 A.M., Mag. c. (r).

— FOLDS, formication, Arg.; itching, and after scratching, burning soreness in afternoon, Arg ; LOWER FOLDS, stitches, > boring with finger, Coloc. (r.).

— INTERNAL, CICATRIX, red, Sabi. (l.); periodical ITCHING, Ant. cr.; PIMPLE, Na. m. (l.); sore on pressure, Apis (l.); suppurating, Pso.; REDNESS, Ant. cr.; SWELLING, Ant. cr.; dull TEARINGS, Upa. (l.).

— LOWER HALF, tearing pressure, Bell. (r.); tension, Thu.

— MARGIN, ACHING when lying, Ple.; BURNING, Cala.; evening in bed, Caus.; DRAWING, Asaf. (r.); INFLAMMATION, Sil.; MOISTURE, Sil.; PAIN, Spig. (l.); REDNESS, Arn.; STICKING, Elat. (l.); < evening, Caus.; SWELLING, Arn., TEARING, Bov., Guai. (l.); TENSION, Bov. (r.); about HELIX, itching stitches, > touch, Ant. cr (r.); OUTER PORTION, constriction, Sars.; tearing, Til.; UPPER, cutting, Lach. (l.); tearing, Anac. (r.).

— POSTERIOR PORTION, pinching BURNING, Stap. (l.); CRAMP, Calc. a.; DRAWING, < touch, Coc. c. (l.); ITCHING, Aga. m.; PAIN, Bry.; PIMPLES, Aga. m., pinching, Spig. (r.); PRESSURE, Lyc. (r.); STITCH, on touch, Bry. (r.); tearing, Bell.; sticking, Meny.

— UPPER PART, BURNING, Con. (l.); HEAT spreading over side of head, thence over face, Olnd. (r., then l.); ITCHING, Carb. v.; PIMPLE, Kre (l.); drawing STITCH, Stan. (l.).

HELIX. Pimples, Ind.; pinching, and in lobe, Elap.; ulceration, Grap. (l.).

HOLLOW. Stabs, Alum.

INNER SURFACE hot to touch, Bry.; tickling, > scratching, Ol. an. (r.).

LOBE, ACHING, Pho.

— BITING, Led.; followed by a node, Lach.; corroding, Plat. (l.).

— BOIL, painful, and discharging matter and blood, Na. m. (l.).

— BLEEDING in drops after rubbing, without a sore or pimple, Sil.

LOBE, BRUISED SENSATION, Chel. (l.), Crot. h.; with heat, *Kre.*, Lach., Merc.

—BURNING, Am. car. (l.), Arn., Bry., Cap., Carb. a., Carb. a. (r.), Carb. v., Chel. (r.), K. n. (r.), Na. p. (r.), Nitr., Rhu. t., Saba., Sabi. (l.), Sars.; WITH redness, Sabi.; twitching, inflammation, and swelling, NITR. (r.); TEARING, Carb. v. (l.).

—CRAMP, Zn. (l.); extending into neck, on boring in finger, Zn. (l.).

—DARTINGS, twisting, Ton.

—DRAWING, Ars., Dro. (r.), Pho., Sars.; with pulling, Sars.

—EROSION, Arg.; as from a caustic, wants rubbing, Plat.

—ERUPTION, Apis, BAR. C., *K. ca.*, Sars.; dry, Elap.; like herpes, Merc. (r.); reddish and rough, Apis (l.).

—FREEZING, in slight cold, Zn.

—HEAT. ACON., *Alum.*, ANGU., Arn., *Bry.*, *Camph.*, Carb. a., CAUS., Chin., CINCH., Hyos., K. ca., Kre., Merc. sol., Na. m., Olnd., *Saba.*, Sabi., Sil., Sars.; EVENINGS, Sil.; WITH bruised sensation, *Kre.*, Lach., Merc.; inflammation, Nitr.; peevish and lachrymose mood, Alum.; heat of head, SIL.; ALTERNATELY in right and left, spreading over same side, and lastly over whole face, Olnd.; BURNING, Rhu. t.

—INFLAMMATION, Aga. m., Alum., Ambra, Angu., Apis, *Arg.*, *Arn.*, Ars., BAR. C., Bro., *Bry.*, Calc. ostr., *Camph.*, Carb. a., Carb. v., CAUS., Cham., Chel., Cic. v., CINCH., Crot. h., Colch., Dros., Grap., Hell., Hyos., K. bi., *K. ca.*, K. n. (r.), KRE., Lach., Lau., *Mar.*, Merl., Merc., Mur. ac., Na. c., *Na. m.*, NITR., *Nit ac.*, Olnd., Pho., *Pho. ac.*, Plat., Pb., Pso., Rhu. t., *Saba.*, Sabi., *Sars.*, *Sep.*, Sil., *Stan.*, Stry., Tab., Thu., Ton., Verat., Zn.; after having been pierced, or when the rings have been torn out of the ears, splitting and tearing the lobe, *Nitr.* (high); with swelling, burning, and twitching, NITR. (r.).

—ITCHING, Aga. m., Alum., ARG., Ars., Asc. t. (r.), Bro., Caus., Grap., K. ca., K. n. (r.), Lau. (l.), Na. m., Na. p. (r.), Pers., Pho. ac., Rhu. t. (r.), Saba., Sars., Verat.; MORNING after rising, Arg.; at NIGHT on washing, K. bi. (r.); (AMEL.) scratching, K. ca. (l.), Lau. (l.), Na. c.; pressure and scratching, Na. m., Pho. ac. (r.), Saba., Sars., Verat.; WITH itching on cheek, Grap.; STICKING, Pho. ac. (r.); as from a TETTER, Caus.; with white, dry scales, Mar. (r.); TICKLING, Aga. m. (l.).

—LYMPH exudes after scratching, Grap.

—NODULES, Merc.; preceded by biting, Lach.; of the size of a lentil, sore to touch, Nit. ac.

—PAIN, Carb. v., Chel., Dro., K. clc. (r.), Merc. sol., Mur. ac., Pho., Zn.; cramp-like, < boring in ear with finger, the same down neck, Zn.

—PIMPLES, Lach., Merc. sol.; burning, corroding, itching, moist, with a scaly look, Merc. sol. (r.); painful, Merc. sol.; lasting twelve weeks, Merl.

—PIERCING, Stan.

—PINCHING, and in helix, Elap.

—PRESSURE, Pip. m. (l.); > evening, Pho.

—PRICKLING, Rhu. t.

—PULLING, Sars.

—REDNESS, Caj., *Camph.*, Cap., *Chin.*, Cinch., Cit. v., K. ca., K. n., Merc. sol., Nitr.; WITH burning, Sabi.; heat, Camph.; redness of cheek, Cinch.

LOBE, SCAB, burning and itching, Sars.

—SCALES, Mar.; dry, white, with itching, Mar. (r.).

—SHOOTING, Kre., Na. m., Pho., Pho. ac., Pb., Saba., Tab., Zn.; with stinging, Saba.; extremely painful and long lasting, Pso.

—SORENESS ON TOUCH, Mur. ac., Nit. ac., Pho.; in evening, Pho.; > evening, Pho. (r.); as if it would ulcerate, Mur. ac.

—STITCHES, Carb. a., Lach., Na. c. (l.), Na. m., Pho., Pb., Pso. (l.), *Saba.*, Saba. (r.), Tab., Ton., Zn.; WITH shooting, Saba.; FINE, Tab. (r.); ITCHING, Na. c., Na. m. (r.); forenoon, > rubbing and pressure, Na. c. (l.); THROBBING, Pho.

—SWELLING, Cit. v., K. n. (r.), Rhu t., *Rhu. t.* (l.), (r.).

—TEARING, Ars., *Ambra*, Canth., Carb a., Carb. v, Cham. (r.), Chin., Cic. v., Cinch., Cup., Guai., Lau., Mur. ac (l.), (r.), Pho. (r.), Stan., Tab. (r.), Verat., Zn.; DRAWING, Ars. (l.); FINE, Tab. (r.); PINCHING, Stan.; TWITCHING, Pho. ac.; evening and night, Na. slfc. (l.); VIOLENT, Ambra (l.).

—TENSION, Thu.

—TETTER, Sep.

—TICKLING, > scratching, Bro. (l.).

—TUMOR, encysted, Nit. ac. (l.); sore on touch, Pers. (r.).

—TWITCHING, Nitr.; WITH burning, inflammation and swelling, NITR. (r.); FINE, Pho. ac.; VISIBLE, Sars.

—ULCERATION, in hole for earring, Stan.; as if beaten, Chel., Crot. h., Lach., Merc.

—VESICLES, caused by discharge from ear, TEL. (l.); small, rough, red, herpetic, Apis.

—WIND, sensation of cold, Stan.

—BEFORE, pain, Buf. s.

—BEHIND, ERUPTION, and on neck, Sep.; ITCHING, Ment. pi. (r.); large NODES, painless, with a white pimple on the top, Stap.; PIMPLE, sore, < touch, Pho. ac. (r.); SORENESS on pressure, Mag. c. (r.); acute STITCH, > pressure, Na. c. (r.); pulsating, Pho.; TEARING, *Ambra*; BONE, pulsation and ulceration, > pressure, Na. c. (l.); periosteum, drawing, into cheek and to lower jaw, Arg. (r.); FOSSA, drawing, Arg. (l.); extending downward in a crescent from, during rest, Arg. (r.); pimple with white tip, Stap.;, pressure, Hell.; stitches into head, Arg. (r.); inward, Arg. (l.).

—CARTILAGE, griping tearing, Stan. (l.).

—INNER SURFACE, ITCHING, Arg.; burning, Saba.

—POSTERIOR surface, DESQUAMATION on scratching, Mez. (l.); burning ITCHING, Mez. (l.); PIMPLE, sore on touch, Nit. ac.; TENSION, Mez. (l.).

LOWER CORNER. Hard and painful swelling, Pip. m. (l.).

LOWER PART OF. ITCHING, Rhu. v. (l.); PAIN, Arun. (l.); acute, extending to cheek and neck, Am. br.; SORENESS on touch, Thu. (r.); SWELLING, Thu. (r.).

MARGIN. BURNING, Sabi. (l.); ITCHING, Ol. an. (r.); 4 P.M., > rubbing, Ol. an. (r.); PIMPLE, burning, sticking, bleeding, after rubbing, with itching, > touch, *Na. m.* (r. and l.); SCALES, Bry.; SORENESS, Bry.; TEARING, Til. (r.); TICKLING, Bro. (r.); ANTERIOR MARGIN, burning, Alum. (r.); tearing, K. ca. (r.); SKIN OF MARGIN, dry, thickened, and white, Bry.

MEATUS, ACHING. Na. p. (r.), TEL. (l.), Thu., Verat.

—AIR (including WIND), SENSATION of, evening, Mez. (r.); < yawning, > boring in, Mez. (r.); as if FREE ACCESS WERE PREVENTED, forenoon, Thu. (r.); air ENTERING, Amph.; when opening and shutting mouth, Thu.; with desire to bore in, Mez.; with stitches to the eyes, Pul.; COLD AIR, Dul., Plat.; rushing in, Lachn.; out when laughing, Mil. (l.); RUSHING out, Bell., Canth., Chel., Pso. (l.), Rhu. t. (r.), Stan., Stram.; after whistling and ringing, Vinc. mim.; hot, Æth., Par.

—BLISTERS, Nicc. (l.).

—BORING, Merc. i. f. (r.), Sul., Upa. (r.); evening, Ran. s. (l.); DESIRE for, Aga. m. (l.), Arun., Colch., Mez. (r.); in evening, Physo.; with feeling as if the ears were too open, and air were rushing, or as if tympanum were exposed to cold air, Mez.; PAIN, later bruised pain from pressure, Caus. (r.).

—BURNING, Arun.; itching, Mag. m.

—CLOSED SENSATION, > boring finger in, Spig.

—COLDNESS, Mez. (r.); with moisture, Merc. sol.; in a small spot, Chr. ox. (l.); as from a wind, Caus. (r.), Stap. (r.).

—COMPRESSION, Asaf.

—CONSTRICTION, < removing wax, Bry.

—CONTRACTION, Bry.; cramp-like, Anac. (l.).

—CRAMP, Anac., Fer. mu.; < drawing scalp down from highest point of skull, Thu. (r.).

—CRAWLING, > boring in, Mil. (l.).

—DENUDED, *Merc. sol.* (< r.).

—DILATATION, SENSATION of, Mez. (r.); MORNING on putting in finger, Mez. (r.); in EVENING, Mez. (r.); < yawning, > boring finger in, Mez. (r.); by AIR, MEZ. (r. and l.).

MEATUS, DISCHARGE. Æs., Agn., ALL. C., Aloe, *Alum.*, Alum. (r.), Ambra, AM. CAR., Am. m., ANAC., Ant t., Apis, *Ars.*, Ars. (l.), Arg. n., Ars. i., *Asaf.*, AUR., *Bap.*, *Bell.*, Benz. ac., Berb., *Bor.*, Bor. (l.), Brach., Bro., BOV., *Bry.*, *Cact.*, CALC. OSTR., Calc. ostr. (r.), Calc., Calc. (l.), Carb. a., Carb. a. (r.), *Carb. v.*, *Caus.*, *Cham.*, Chim., CIC. V., Cina., *Cist.*, COLCH., *Con.*, Croc., *Elap.*, Elap. (l.), Eup. p., *Gel.*, *Grap.*, *Hep.*, *Hydrs.*, *Iod.*, *Ir. v.*, K. BI., *K. ca.*, KIN., Kre., LACH., *Lachn.*, Lith., LYC., MENY., MERC., *Merc. c.*, *Merc. sol.*, Murx., NA. M., *Mos.*, NIT. AC., *Petrol.*, PHO., PHO. AC., Phyt., Pso., PUL., *Pul.* (l.), Pul. n., Rum. c, RHU. T., *Sang.*, SEL., Sen., SEP., SIL., *Spig.*, Stil., SUL., Sul. (l.), TEL., Tel. (l.), *Thu.*, Thu. (r.), *Vac.*, Verat. v., ZN., Zn. (l.), Xan.; in AFTERNOON, Bry.; NIGHT, Sep. (r.); in warm bed, Merc. sol.; after acute ERUPTION, Meny.; after ITCH, Carb. v.; after MEASLES, Colch., Meny.; after abuse of MERCURY, Asaf., Aur.; itching on OCCIPUT, Bor.; after SCARLATINA, K. bi., Lyc., Meny.; after SPATTERING, Spig; supposed to be from VACCINATION, Vac.; CHILDREN are better when it runs, worse if not, Sul.; with delicate white skin, Caus.; in several young persons, Sul.; with throbbing in ears, Tel.; with coldness, Merc. sol.; caries of mastoid process and ossicula, Aur., Fl. ac.; inflammation of external meatus, Caus., K. ca., Sep., Sil., Sul., and of membrana tympani, Carb. v.; sensitiveness to touch, Carb. v., Cham.; swelling of outer ear, Bor., Cist.; itching in ears, Anac., Bor., Merc., Petrol., Sep.; and eruption, Sul.; itching and swelling, Tel.; swelling, Caus., Cist.; external swelling, Sil.; bluish red color of ear, and looking as if infiltrated with water, Tel.; ulceration of inner ear, Lyc., of outer ear, Merc.; tearing pain, Colch., Merc., after measles, Colch.; with pain from ear down neck when turning head, Carb. v.; noise in ears, Calc. ostr.; roaring, Bor.; hard hearing, Am. m., Asaf., Calc. ostr., Carb. v.; Caus., Elap., Lyc., Sil., Tel.; with HEADACHE, Pso.; shooting in head, Bor., in forehead, Elap.; burning pain on outer head, extending down neck, Carb. v.; flow of TEARS, Elap.; eruption on FACE, Sul.; vesicles, Merc.; paralysis of face, Caus.; large ABDOMEN, Calc. ostr.; swollen glands in NECK, Calc. ostr.; vesicular eruption on neck when the discharge touches the skin, Tel.; pustules on lower LIMBS, Merc. s.; little warts on hands and fingers, Calc. ostr.; swelling of knee, Sil.; PROSTRATION and sinking, Ars.; desire to be UNCOVERED, Lyc., Pul., Spig., Sul.

MEATUS, DISCHARGE, FOLLOWED by hard hearing, Bor.

— —BLOODY, Am. car., Arun. (l.), BELL., *Bry.*, *Calc. ostr.*, Caus., *Cic. v.*, CON., *Crot. h.*, Elap., Ery. a. (l.), *Grap.*, Ham. (r.), Lach., LYC., *Merc.*, MERC. s. (r.), Mos., *Nit. ac.*, PETROL., Petrol. (< l.), PHO., *Pul.*, Rhu. t., SEP., SIL., *Sul.*, Zn.; in DROPS, Mos. (r.); after a sound as of a cannon, Mos.; OOZING, Crot. h., *Pho.*; morning, Merc. s. (l.); RUNNING from ear, BELL., BRY., CALC. S., CIC. V., CON., Cic. v., *Grap.*, LACH., LYC., MERC., *Mos.*, NIT. AC., *Petrol.*, PHO., PUL., *Rhu. t.*, SEP., SIL., SUL., *Zn.*; suddenly, *Crot. h.*; if the ears run blood, the child makes less water, Carb. caus., Colch., Lyc., Merc., *Pho.*, Pul.; SPURTING, Cary., Cic. v.; of arterial, Elap.; and from nose, Elap.; AND PURULENT, Cannab. s. (r.), Caus., Ery. a. (l.), Merc. sol. (r.), Petrol., Rhu. t.

— —BROWN, Anac., Tarent. (r.); thick, Carb. v.

— —CADAVEROUS, Ars., Thu.

— —CATARRHAL, every seventh day, Sul.

— —CLEAR, BRY.

— —CORRODING, Ars., Calc. ostr., Calc. ph., Hep., Lyc., Merc., Sul., *Tel.*, Tel. (l.); causing eruption, intertrigo behind ear, itching and bleeding after scratching, Sul.

— —FLESH-COLORED, offensive, *Carb. v.* (r.), K. ca., Zn. (l.).

— —FLUID, see WATERY.

— —GREEN, after scarlet-fever, without pain, Bov.; yellowish, in morning, Elap.

— —LYMPH, oozing of, with itching in ear, Grap.

— —MILD, not corroding, and without smell, Pho.

— —MUCOUS, *Alum.*, *Bell.*, Bon., *Bor.*, CALC. OSTR., *Grap.*, LYC., MERC., PHO., PUL., SUL., Tarent (r.); fetid, Calc. ostr., Lach., Sul.

— —OFFENSIVE, Ars., Asaf., Aur., Bov., Calc. ostr., *Carb. v.*, *Carb. v.* (r.), Caus., Cist., Ery. a. (l.), Grap., HEP., Hyos., K. ca., Lach., Lyc., *Merc.*, Merc. c., MERC. SOL. (r.), Nup., Pso., Sep., Sul., Tel., Thu., Zn., Zn. (l.); with itching and deafness, Bov.; with itching in ear, and swelling of lobe, Tel.

— —PURULENT, Acon. (l.), Æth., *Alum.*, Alum. (r.), All. c., AM. CAR., Am. m., Arun., Ars., ASAF., AUR., BELL., BOR., Bor. (r.), Bov., Calc. c., CALC. OSTR., *Carb. a.*, CARB. V., CAUS.. CHAM., Cist., CON., Cop. (l.), Gel., GRAP., HEP., Jug. r. (< l.), K. bi., K. CA., LACH., *Lyc.*, MERC., NA.

M., NIT. AC., PETROL., *Pho.*, PSO., PUL., RHU. T., Sac., *Sep.*, SIL., SUL., Tep., ZN., Zn. (l.); DAY and NIGHT, Zn.; WITH hard hearing, Asaf., Bor., Pul., at times, Sul.; inflammation of external and internal ear, Bell. (r.); lancinating pains, Bell., Cham., Chin.; otitis, Bell.; swelling in meatus and otitis, Pul.; AND BLOODY, Cannab. s. (r.), Caus., Ery. a. (l.), Merc. sol. (r.), Petrol., Rhu. t.; with burning in ear, Pul.; with burning pain, Chin., Pul.; and BROWNISH, Anac.; and FETID, Cist., Merc., MERC. SOL. (r.), Pso., Sep.; with hard hearing, Asaf., Aur., Bov.; PROFUSE, with cadaverous odor, Ars.; and THICK and yellow, from both ears, after scarlet-fever, K. bi.; and WHITE, Ery. a. (l.); and YELLOW, Merc. sol. (l.); SMELLING, Bry.

MEATUS, DISCHARGE, THICK, *Carb. v.* (r.), Ery. a. (l.), Tarent. (l.); and brown, Carb. v., Carb.v. (r.).

— —WATERY (including FLUID, THIN, etc.), ASAF., BELL., *Calc. ostr.*, CARB. A., CAUS., Cist., Elap., Elap. (l.), *Kre.*, MENY., MERC., NA. M., NIT. AC., *Pho.*, Sep., SIL., *Spig.*, Tarent. (r.), Tel. (l.); in MORNING, *Elap.;* and CADAVEROUS, Ars.; DRIPPING, Rhu. t.; smelling like FISH-PICKLE, Tel.; oozing, like PUTRID MEAT, Thu.; and YELLOWISH, K. slfc.; WITH blotches in throat, Elap.

— —WHITE and purulent, Ery. a. (l.).

— —YELLOW, MERC., Na. m., Pho.; greenish, Elap., Gel., Pul.; and watery, K. slfc.

— —SENSATION OF, Aga. m., Merc., Sil. (l.); at night, Bry.; with drawing pain, Mil. (l.); of water, Acon. (l.), Calc. c., Chr. ac. (r.), Cinnb., Der., Grap. (l.), Mil. (l.), Merc. sol., TEL. (l.); after dinner, Thu. (r.); cold water, Merc. sol.; tenacious liquid, Na. m.; sensation as if about to discharge, Lachn.; afternoon and evening, Hip. (l.).

MEATUS, DRAWING. Anac. (l.), Asaf., Chel. (l.), Dul., Nit. ac., Sil.; in EVENING, Coc. c. (r.), Ran. s. (l.); with sensation of a discharge, Mil. (l.); CRAMP-LIKE, Croc.; EXTENDING backward, noon, Aloe (r.); outward, Sul. ac. (r.); into temple, Chel. (l.); SUDDEN, Coc. c. (r.).

—DRYNESS, see WAX, WANT OF.

—EXCRESCENCE, fungous, Merc.

—FLEA in, sensation of, Hæm.

—FLY in, sensation of, Elap.

—FOREIGN BODY in, sensation of, Canc. f. (r.), Pho.; before drum, Calc. a.

—FORMICATION, Ant. cr. (r.), Sul. (l.); > boring in with finger, Mil. (l.), Ton. (r.); biting, Plat. (r.); deep-seated, Ars., Lau.

—FULNESS, Cinnb. (l.).

—GNAWING, Sul. (l.).

—HEAT, Asar. (r.), Chel. (r.); rushing in, Ant. cr., Lyc.; rushing out, Æth., Calc. c. (l.), Cle., K. ca., Ol. an., Par.; after two beats in ear, Ol. an.

—HERPES, habitual, with hard hearing, Grap.

—INFLAMMATION, Arun., Mag. c. (r.), Petrol.; WITH soreness, Merc.; swelling, Calc. ostr., Cist., K. ca., Na. m., Sep., Tel., Thu., Zn. a.

—ITCHING, Aga. m. (r.), (l.), Alum., Arun., Bov., Coc. c. (l.), (r.), Elap. (r.), Fago. (l.), Fer. mu., Fl. ac., *Ig.*, K. n., Lau., Mag. c. (r.), Manc. (r.), Merc. d., Merc. i. r. (r.), Mil. (r.), Na. p. (r.), Ol. an., Phe., Sars. (l.), Sil., Sul.

(l.), Zn.; in FORENOON, Fago.; 2 P.M., Fago. (r.); EVENING, Elap.; after oozing of LYMPH, Grap.; (AMEL.) boring in with finger, Bov., Coc. c. (l.), Fl. ac., Lau., Mil. (r.), Ol. an., Phe., Zn. (r.); scratching, Mag. c. (r.); ACUTE, > touch, Hypc. (r.); alternately in one or other, Chel.; child BORES in, Fl. ac., Mez.; ITCHING of ears internally, Dios.; in ear internally and externally, Spira.; inside ear in evening, Elap., Murx.; of ear internally, 10.30 P.M., Dios. (r.), (l.); BURNING, Arun.; DEEP-SEATED, 8 P.M., Rum. c. (r.); EXTENDING deeper on boring in with finger, Phe.; to interior of cheek, in course of Steno's duct, Elap.; TICKLING, K. n. (r.); > boring with finger, Aga. m. (l.); VOLUPTUOUS, extending through inner ear to mouth, Coc. c.

MEATUS, LANCINATION, Ast., Crot. c.

—MEMBRANE stretched across, sensation of, Asar., Asar. (r.); < cold weather, Asar.

—PAIN, Abs. (l.), Aloe (l.), Apoc. c., Arun., *Asaf.*, Chel (l.), Cinnb. (l.), Hæm., Merc. i. f. (r.), Spira., Sum.; in EVENING, Caus. (r.); on BORING in with finger, Rhodo. (l.); on TOUCHING, Tab, Zn. ac. (l.); < PRESSING TEETH TOGETHER, Aloe (l.); ACUTE, Merc. i. f. (r.); SPASMODIC, Anac.; SUDDEN, Merc. i. f. (r.).

—PICKING in, Bov. (l.), Dro., Dro. (r.).

—PIMPLE, Jug. r.

—POLYPUS, CALC. OSTR., Dul., Merc., Stap., Thu.; stinking, tincture Calc. in water, lime-water.

—PRESSURE, Asaf. (r.), Bell., Chel. (r.), Coc. c., Rhe., Sil.; WITH tension, extending to left lower jaw, and salivation on right side, Asar.; DRAWING, Bism. (l.); EXTENDING, to right lower jaw, *Asar.* (r.); as from a FINGER, < stooping when reading, Bry.; INWARD, Spig.; STICKING, Nx. v.; TEARING, Sars. (r.); TENSIVE, < cold weather, Asar.; TOWARDS, Op.; against TYMPANUM, Anac. (l.).

—PULSATION, TEL. (l.).

—PUSTULES, Pte. (r.); sore, Cannab. s. (r.), Gas (l.).

—REDNESS, Mag. c. (r.); with swelling, stitching, and tearing in ear, and catarrhal affection of Eustachian tube, Pul.

—RELAXED SENSATION, in morning on putting in finger, Mez. (r.).

—RUSHING out of something warm, Sul. ac.

—SHOOTING, Bell.; evening in open air, Sul. (r.); cool, Fer. mu.

—SORENESS, Fago. (l.), K. bi. (l.), Merc. sol. (< r.); on CLEANING ear, Caus.; on PRESSURE, Mag. c. (r.); on TOUCH, Fago. (l.), Na. m.; in a SPOT, Sel. (l.); WITH inflammation, Merc.

—SPASMODIC SENSATION, Cham.

—STENCH without discharge, Aur., Bov., CARB. V., Caus., Cist., *Grap.*, Hep., Hyos., *Merc.*, Pso., Zn.

—STITCHES, Acon. (r.), Aga. m. (l.), Angu., Arun., Bry. (r.), Carbn. s., Carb. v. (r.), Cham., Chel. (r.), (l.), Crot. c., *Dul.*, K. bi. (r.), Pso., Ran. s. (r.), Tarent. (r.); NOON, sitting, Gel.; AFTERNOON, after descending stairs, Chin. s. (l.); EVENING, during rest, Pso.; when chewing, Cannab. s.; > BORING with finger, Pso.; WITH narrow feeling, Lyc.; stinging, Camph. (l.); ACUTE, Cham. (r.), Pho. (r.); BITING, Coc. c. (r.); DRAW-

ING, extending outward, Calc. caus. (l.), over outer portion, Krc. (l.); DULL, Plat. (r.); EXTENDING inward, Carb. v. (l.); at 5 P.M., while walking, K. bi. (r.); outward, at night, *Ars.* (l.); FINE, Na. m. (r.); INTERMITTING, Plat. (r.); PRESSING, Thu. (r.); TICKLING, Wies.; deep-seated, Dro. (r.).

MEATUS, STOPPED SENSATION, from without, Ars. (l.); by a swelling, with cramp-like sticking internally and externally, MERC. SOL.

—SUPPURATION, feeling of, Mag. c. (r.).

—SWEAT, Sul. (l.).

—SWELLING, Bry., Calc. c., Caus., Cannab. s., Na. m., *Petrol.;* WITH inflammation, Calc. ostr., Cist., K. ca., Na. m., Sep., Tel., Thu., Zn. a., and purulent discharge, Pul.; pain, Mez. (r.), Petrol., on touch, Zn. ac. (l.); redness, Zn. ac. (l.); redness, stitching, and tearing in ear, and catarrhal affection of Eustachian tube, Pul.; SENSATION of, Junc.; in evening, Mez. (l.).

—TEARING, Canth. (r.), Chel. (r.), Chin., Colch. (r.), Indg. (l.), K. n. (r.), Lyc. (r. and l.), Stan. (r.), Tarax., Ton.; EXTENDING to upper jaw, caused and < cold air, Aga. m. (r.); FINE, Phel. (r.); INTERMITTENT, Pso.; PRESSIVE, Aur. (l.); STICKING outward, < towards evening, Ars. (l.).

—TENSION, *Asaf.* (r.), K. n. (r.); with pressure, extending to left lower jaw, and salivation on right side, Asar.; pressive, *Asar.*

—TICKLING, Coc. c. (l.), Na. c. (l.), Rhodo. (l.); extending through inner ear to mouth, Coc. c.; at 8.30 A.M., Na. c. (l.); > scratching, Na. c. (l.); voluptuous, Ars.

—TINGLING, Alum.

—TWINGING, Anac. (r.); in evening, Mez. (l.).

—TWITCHING, Anac. (l.), Nit. ac., Val. (r.); tearing, Carb. v. (r.).

—ULCERS, Alum., Bov. (r.), (Bry.), *Camph.*, Grap. (l.), Kali., *Merc.*, *Pul.*, *Rut.*, Spo., Stan.; WITH discharge, Lyc.; pain on swallowing, Bov. (r.); RED, with sticking pressure on touch, *Camph.;* SENSATION of, on cleaning ear, Caus.; on putting on finger, Sep.

—VAPOR, sensation of a hot, coming from, Canth., Par.; going in, Euphr. (r.).

—WATER rushing in, sensation of, with a large quantity of thick or thin wax, Lyc.

—WAX, ABUNDANT (including INCREASED), Aga. m., Am. m., Bell., Calc. c., CALC. OSTR., *Carb. v.*, Cham., Coca (r.), Colch., CON., Cyc., Elap., Grap., Hep., Iod., *K. ca.*, *Lach.*, Lyc., *Merc.*, Merc. i. r., Mos., Mos. (l.), Mur. ac., Nit. ac., Petrol., Pho., *Sed.*, SEL., Sel. (l.), Sep., Sil., Sul. (< l.), Tarent. (r.), Tel., Thu., Wies., Zn., Zn. (l.); with gurgling, K. ca.; with dull hearing, Mur. ac., Zn. o.; with itching, in forenoon, Cyc., Mur. ac., Sep.; with roaring, Mur. ac., Sep.; with rushing as of water, Lyc., Petrol.; with swelling of outer ear, Calc. ostr., Na. m.; in BALLS, Dios. (r.), Elap.; BLACK, Elap.; and hardened, Elap., Pul.; with hard hearing, Pul.; dark BROWN, in evening, Calc. s. (r.); DARK and hard, Mur. ac.; DRY, Æth., Carb. v., Cast. eq., Cham., Colch., *Grap.*, Lach., Mur. ac., Nit. ac., Petrol., Pho.; with buzzing, Pho.; with hard hearing, Lach.; with swelling of ear, Nit. ac.; with or without rushing, Pho. ac.; with hard hearing (after Sul.), Lach.; FLOWING out, Wies.; at night, Mos. (r.); with itching, Am. m., Anac., Petrol.; roaring, Grap.; flowing in Eustachian tube, Crot. h.,

Lach.; FLUID, Am. m., K. ca., Merc. s., Sil., Sul., Tel., Wies., Zn. o.; running out, Am. m., *Con.*, Iod., K. CA., LACH., Merc., Mos. (r.), SEL.; HARD, Elap., Sel., Sel. (l.); with hard hearing, Sel.; MUSHY, Chel.; like rye-mush, pap, Lach.; PALE, Wies.; like chewed PAPER, with hard hearing, Con.; like shreds of mouldy paper, *Con.;* PURULENT, Sep.; REDDISH, Pso. (l.); blood-red, *Con.;* brown red, dry and hard, Mur. ac.; SLIMY, Wies.; soft, Sil. (r.), Wies.; THICK, Chel.; VISCID, Sul.; WANT OF, Æth., Alco., Berb. (l.), Bon., Calc. ostr., *Carb. v.*, Castor., Cham., Grap., Iod., Lach., Mur. ac., Nit. ac., Petrol., Pho.; with beating, Berb.; with inclination to bore in, Colch.; with bubbling, Berb.; with hard hearing, GRAP., Mur. ac.; WHITISH, Chel., Sep.; accumulates, with itching, Sep.; YELLOW, K. ca.; fluid, SIL.; SENSATION AS IF INCREASED, Aga. m., Calc. ostr., CON., Cyc., Hep., Petrol., Sel., Sep., Sil., Thu.; sensation as if flowing, Aga. m.; SENSATION AS IF IT WOULD FLOW into mouth, Crot. h.; would flow out, on swallowing saliva, Coc. c. (l.).

MEATUS, WEDGE driven in, sensation of, evening while walking in open air, > boring with finger, Par. (<r.).

—WIND, see AIR.

—YAWNING, sensation as from, Olnd.

—ANTITRAGUS, NEAR, stitches on pressure, Sep.

—BEFORE, CONTRACTION, Cham.; sticking ITCHING, Caus. (r.); STITCHES, Cham.; TEARING and TWITCHING, Dro. (l.).

—CARTILAGE, tickling, morning in bed, K. n. (r.).

—ENTRANCE, BURNING, Olnd. (l.), Spo. (r.); DRAGGING towards, Crot. t.; ERUPTION, burning and itching, Rhu. t.; HEAT, Zn. (l.); ITCHING, Caus., in bed, Verat. v. (l.); PAIN, Hype. (r.); < touch, Tarent.; feeling as if a NAIL were driven through head, < touch, Tarent.; inflamed NODE, Spo. (l.); SWELLING, Sep., Zn. (l.); inflamed, K. bi. (l); with tension and crawling, Spo.; pleasant TICKLING, and on wings of nose, alternating with crawling and itching at anus, Saba.; TWISTING towards, Crot. t.; IN REGION OF, pressure and tension, Asar.

—EXTERNALLY, ABSCESS, with right-sided headache, Merc.; pain as if AIR were rushing in, Amph.; INFLAMMATION, Acon., Pho. ac.; with discharge, Caus., K. ca. Sep., Sil., Sul.; ITCHING, Caus., *Ig.*, Nitr.; MOISTURE, with hard hearing, Merc.; NARROWING, Mang.; SUPPURATION, with inflammation of ear, Kin.; SWELLING with inflammation, K. bi. (l).

—INTERNALLY, ACHING, Cup. ar. (r.); at night, in bed, Merc. i. f. (r.); BORING, acute, intermittent, Merc. i. f. (r.); ITCHING, Lob. s. (l.); PAIN, Cup. ar (r.), Hydrc. (l.), Mez. (r.); PRESSURE, intermittent, tearing, Chel. (r.); SWELLING, painful, Mez. (r.); TEARING, Ammc. (r.), Chel. (r.).

—LOWER PART, painful swelling, Mez. (r.).

—CLOSE TO MEATUS, squeezing pain, Sul. (l., then r.).

MEMBRANA TYMPANI. INFLAMMATION with discharge, Carb. v.; violent drawing shooting PAIN from stomach to, every time he laughs, Mang.; PRESSURE inward upon drum, Vio. t.; ULCERATION, thickening and destruction, Hep., Merc.

MUSCLE. LEVATOR AURIS, tickling, > scratching, Bro. (r.).

POSTERIOR PART, HEAT, Aloe., ITCHING, Mos.; external PULSATION, Aloe;

ROUGHNESS, Rhu. v.; sensation of SWELLING, on turning head to left, Rap.; VESICLES exuding yellowish serum, Rhu. v.

SIDE, biting, Lyc. (r.); lancinations, Tarent. (r.); soreness, Lyc. (r.).

SKIN, swelling, Rhu. v.

SKULL, at union with, sensitiveness, Bry.

TIPS, burning, Chel., Coloc. (l.); while the tip of the nose is cold, Chel.; livid, Op.; tearing in forenoon, Castor. (l.).

TOP, drawing towards occiput, Sul. (l.); burning stitching, Calc. s.; pimples, Calc. s. (r.).

TRAGUS. BOIL, Sul.; ERUPTION, scurfy, with burning biting, Pul.; neuralgic PAIN, evening, Fago.; STITCHES, extending into meatus, Cham. (r.); TEARING, Nit. ac.; INSIDE, itching, Mur. ac. (l.); > scratching, Chin. s. (l.); smarting pain, Mur. ac. (l.); BEFORE region of tragus, spasmodic sensation sometimes into meatus, *Cham.*; stitches, Cham. (r.); BELOW, smarting pain and itching pimples, Mur. ac. (l.).

UPPER PART, burning, Apis (l.); pressure as with a plug, Bro. (l.); soreness in evening on pressure, Mez.

ABOUT EXTERNAL EAR. ACHING internally, Bro. (l., then r.); BLISTERS with swelling of external ear, Ars.; BOILS, Am. car.; BORING, Am. m. (l.), Bell. (r.); BRUISED sensation extending down neck to clavicular and scapular region, Coc. c. (l.); BURNING, Calc. c.; COLDNESS, as if in bones, Æth. (r.); DRAWING, Asaf. (l.), Nit. ac. (l.); towards evening, Cle. (r.); upward and downward in a spot, > pressure, Grat. (l.); FULNESS, and in ears, Glo.; HEAT, in evening, Ment. pi. (l.), and burning with hard hearing, Jac.; HERPES and ulcers, and on external ear, Calc. p.; ITCHING, in evening, Ment. pi. (l.); > rubbing, Phel. (l.); NUMB sensation, with hard hearing (after Sul.), Lach.; PAIN, Bry. (l.), Canc. f. (r.), K. ca., Merc. i. f. (r.); after a walk, Pal. (r.); burning, Calc. ostr.; extending upward, Ol. an., Sars.; PIMPLES, (Ant cr.), Mag., Ment. pi. (l.), Mur. ac., Petrol.; itching, Na. p. (r.); feeling as if pimples would form, Tet. (l., then r.); PINCHING, extending towards eye, Glo. (r.); PRESSURE, Æs. h. (r.); REDNESS, Arn.; SENSITIVENESS, and of ear with inflammation, Merc.; SHOOTING extending to occiput, 7.30 P.M., Fago. (r.); SORENESS as if in bone, Tet.; SPASMODIC sensation, and in ears, Ran. b.; STITCHES, Asaf. (l.), Lepi. (r.); towards evening, Cle. (r.); extending outward, morning till evening, Na. m.; towards left eye, Glo. (r.); and in ears, Con., Vio. o.; burning SUPPURATING, and on external ear, Cic. v.; SWELLING, Arn., For.; with otitis, Pul.; TEARING, Canth. (r.), Ton.; bruised, Ery. a.; extending upward and downward, in a spot, > pressure, Grat. (l.); fine sticking, extending to top of head, Æth. (l.); tearing with stitches, and in head, Con.; TENSION, Am. car. (l.), Asaf., Grap. (l.); in morning, Stry.; with dulness and stupidity, Asar.; TETTERS, and on external ear, extending to meatus, Cist.; TWITCHES, Aga. m., Am. car. (l.); UNCOMFORTABLE, For.; ULCERS, Calc. p.; sensation of WARM WATER flowing from, Cala.; BONES, aching, Nit. ac. (l.); digging, at night, Mang.; drawing, Nit. ac. (l.); pain, Bry.; shooting outward, Calc. p.; sensation of swelling, Acon.; tearing, K. ca. (l.); HEAD, dulness, Æs. h.; bursting sensation when vomiting, Asar.; drawing, Nit. ac. (l.); fulness, < out of doors, Linu.; lancinations at night, Tarent.; sensation of something lying on, Plan.; pain extending from one ear to the other, Chel.; pressure, Bry., extending across vertex to other ear, Nit. ac. (l.); stitches, Pho.; tearing, Pho., Sul.; about lower JAW, drawing, K. bi., Petrol.; stitches, K. bi., Lau. (r); tearing, K. ca.; tension, Petrol.; twinges, Colch. (r.); external THROAT, pimples, Sul. (l.); stitches, Hep.; tearing, < pressure, Gam. (l.).

ABOVE EXTERNAL EAR. ACHING, Dul. (l.), Tel.; 5 P.M., Dios.; superficial, extending to upper margin of concha, Mez. (r.); BALD SPOT, Pho.; sensation of a BAND across, Am. br.; BITING, itching, > scratching, Grat. (r.); BORING, Cannab. i. (r.); BURNING, Apis (l.); COLDNESS, externally, spreading in rays, Indg.; like a stone, evening, Lac. ac.; COMPRESSION, Hur.; during stool, Ox. ac.; CONGESTION, burrowing, Coc. c. (l.); CONSTRICTION, Lach., Murx.; CUTTING, Carb. v. (l.); DRAWING, Asaf., Coloc. (r.), Mez. (r.), Verat. (r.); in bed, Chel. (l.); in an old scar, Lach. (r.); extending to crown, Lach. (r.); ERUPTION, itching, scurfy, *Stap.;* FULNESS, *Glo.;* at noon, Sarr., ITCHING, > touch, Ill. (l.); spreading over whole body, in morning, Am. car.; PAIN, Ced., Chel. (r.), (< r.), Ger. (l.), Hur. (l.), Merc. sol., Plumbg.; in evening, Chin. s. (r.); 8 P.M., Dios.; acute, 9 P.M., Dios. (r.); extending through upper back teeth, Chel. (l.); light, with numbness of jaw, Hur.; pulsating, Lepi. (l.); PIMPLES, red and itching, Cop.; PINCHING, Carb. v. (l.); PRESSURE, Arg. (r.), Camph. (l.), Ced., Dul. (l.), Mez. (l.), Nx. m. (r.); externally, Sabi. (r.); PULSATION, Glo., Lepi. (l.); PUSTULES, containing serum, Sum. (l.); SHOOTING, 6 P.M., Erio. (r.); SORENESS, on touch, Lyc. (l.); STIFFNESS, Plan.; STITCHES, Asaf. (r.), Mur. ac. (r.), Sep.; 4 P.M., Merc. sol. (r.); when walking, Ars.; acute, K. ca.; dull, Mag. c; fine, Plan.; pressive, Coc. c. (l.); spreading in rays, externally, Indg.; TEARING, Camph., Chel. (r.), Na. slfc.; pressive, Arg. (l.); TENSION, in an old scar, Lach. (r.); TREMBLING, Arg.; TWITCHING, Aga. m., K. ca.; BONE, burning, aching, extending inward, Stap. (l.); drawing posteriorly, Mez. (r.); pain, Bry. (l.); pressure, *Pul.* (r); swollen sensation, Plan.; tearing, Led. (r.), Merc.; < touch, Aga. m. (r.); posteriorly, Led. (r.); tension, Coc. c.

BEFORE EXTERNAL EAR. ACHING, Anac. (l.), Cup., Dios., Merc. i. f., Na. m.; 3 P.M., Dios.; extending to angle of jaws, Dios.; BOIL, with suppurating pain on touch, Lau. (r.); BORING, Lau.; on bending trunk to right, Mag. m. (l.); BRUISED sensation, on touch, Zn. (r.); BUBBLING, Lyc.; BURNING, > RUBBING, Grat. (l.); BURSTING sensation, Dios.; COLDNESS externally, extending like rays, Indg.; CRACK extending from left upper lip over cheek to ear, Am. car.; DULNESS, in evening, Cham.; FULNESS, Glo., Lact.; sensation of something HEAVY, Carb. v.; and in ears, with a stopped sensation, Carb. v.; heaviness, and in ears, with hard hearing, CARB. V.; ITCHING, Ol. an.; > scratching, Ol. an.; touch, Ill. (l.); NUMBNESS, Sul.; PAIN, Dios. (r.), Der. (< l.); 3 A.M., Dios.; 8 A.M., from cough, Dios. (r.); 11 A.M., Dios. (l.); acute, morning, Dios.; extending to ear, when raising hand to right side of neck, Elap.; PAPULES, Na. c.; PRESSURE, Dios.; inward, Sep.; PULLING, Dios.; PULSATION, morning, Lyc.; evening, Lyc.; after lying down, Hep.; on lying on ear, Bar. c. (l.); while standing after dinner, Castor. (r.); PUSTULE, Mag. c.; SCREWED IN sensation, Sul. ac. (l.); SORENESS, Senec. (l.); on touch, Pte. (r.); SQUEEZING sensation, extending to angles of jaw, Dios.; STIFFNESS, Plan.; STITCHES, in evening, Ran. s. (r.); acute, Mag. m. (r.);

5.30 P.M., Mag. c. (l.); burning, 2 P.M., Ol. an. (r.); cutting externally, Arg. (l.); extending downward, Verb. (l.); into ear, Stro. (l.); fine, Lau. (l.), Mil. (l.), Plan.; spreading like rays externally, Indg. (r.); tearing, in evening, Con.; SWELLING, Pte. (r.); boil-like, discharging, Bry.; painful, like a boil, on touch, *Calc. c.* (l.); TEARING, Bov. (r.), Mag. m., Ol. an. (l.), Stro. (r.), Tab. (r.); 4 P.M., Mag. c. (l.); after dinner, Bov. (r.); acute, Na. p. (l.); extending into cheek, Sul. ac. (l.); into temple, K. iod. (r.), (l.), Sul. ac. (l.); external, Mag. m. (r.); piercing, Rat. (r.); twitching, externally, Tab. (r.); TWISTING, 11 P.M., Dios. (l.); TWITCHING, Mag. m.; on moving head, Sul. ac. (l.); tearing, Angu. (l.); ULCER, Carb. v.; discharging through ear when touched, *Merc. sol.* (r.); sensation of an ulcer, in morning, Sars. (l.); VERTIGO passing up to vertex like a wave, Sol. p.; VESICLES, painful and filled with pus, Cic. v.; sensation as if he would VOMIT, Dios.; BONES, boring, Aur. m. n. (l.), *Bar. c.* (r.); construction, Zn. (r.); crawling, Bar. c. (r.); drawing, *Bar. c.* (r.); gnawing, Sul. (l.), on swallowing, Sul. (l.); pressure, Aur. m. n. (r.); swollen sensation, Plan.; tearing, Bar. c. (r.), Indg. (r.), Rat. (r.), Zn. (l.); near EYE, tearing, Grat. (l.); SKIN red, rough, tetter-like, with oozing and bad smell from behind the same ear, and sticking from the ear to the head, Olnd. (l.).

BEHIND EXTERNAL EAR. ACHING, Mos., Na. m. (l.), Stry., Thu. (l.), Vio. o. (l.); on going into open air, > touch, Mang. (l.); on shaking head, Glo.; extending to temples, Ced. (r.).

— AGREEABLE SENSATION on bathing in cold water, Fl. ac.

— BITING, Lyc. (r.); > scratching, Ol. an.

— BOIL, Bry., *Caus.*, Na. c., Phyt., Stap., Thu. (r.); tearing in, *Angu.* (r.); boils and tumors, Con., Rum. c.

— BORING, Aur. (l.), Aur. m. n., Cannab. i. (r.), Caus., Caus. (l.), Coloc. (r.), Cup., Mez. (r.), Mos., Rum. c. (l.), Saba.; evening, Ran. s. (r.); while walking, Mez.; behind and in ear, Cup.

— BLISTERS, and on neck, small, burning, painful, in evening, sensitive to pressure next morning, Am. car. (l.).

— BRUISED SENSATION, Chel. (l.), Cic. v. (l.), (r.), Lachn.

— BURNING, Aur., Na. m. (r.), Rhu. v., Thu. (r.), Saba.; 4 P.M., Grat. (r.); night, Aur. m.; in spot, Calc. ph. (r.).

— CLEFT SENSATION, Am. car. (l.).

— COMPRESSION, < walking, > sitting, *Asar.*

— CONTRACTION, Stry.

— CRACK, Chel. (r.); in a downward direction, Chel.

— CRAMP, Murx.; on going into open air, > touch, Mang. (l.).

— CRAWLING, All. c. (r.).

— CRUST, Aur. m.

— CUTTING, Bell., Carb. v. (l.); extending down neck, 10.15 P.M., Ir. fœ.

— DARTING, Stry. (r.), Xan. (r.); evening, Dios. (l.); extending in front of ear and to angle of jaw, morning, Dios. (l.).

— DRAWING, Anac. (l.), As. f. (l.), Coloc. (l.), K. bi., K. ca., Sul. (l.), Thu. (r.); in bed, Chel. (l.); < touch, Sil.; extending to lower jaw, Zn. (l.); to mastoid process, Chin.

— ELEVATION on red base, with burning, stitching, and twitching, Ars. (r.);

small elevations, Bar. c., preceded by itching, and followed by soreness after scratching, Mez.

BEHIND EXTERNAL EAR. ERUPTION, Ant. cr., Canth., Cast. eq., Cinch., Guare., Jug. (r.), K. iod, Olnd., *Pul.*, Saba., Sel., Stap.; and on external ear, Cinch., *Cic. v.*, K. iod., Petrol., *Pho.*, Pul., Sep., Sil., Spo., *Sul.*; ITCHING, Mag. s. (r.); after scratching, Mag. m.; burning itching, < night, Vio. t.; itching of an old eruption, Mag. m.; resembling itch, in children, Arun.; MOIST, Calc. c.; and on external ear, Calc. ostr.; SCABBY eczema with deafness, Pso. (r.); SCURFY, Ant. s. (r.); and on external ear, Hep.; humid, and scurfs on external, Pso.; humid suppurating, and on external ears, Lyc.; itching, Stap.; SORE, Pso. (r.); red irregular SPOTS, Cocc.

—FORMICATION, Bry.

—FULNESS, Thea.

—GNAWING, K. iod.

—HEAT, dry, Merc. sol.; extending to vertex, Pso.

—HERPES, moist, Am. m.

—IRRITATION from pin or any thing, Plan.

—ITCHING, Aur., Calc. c., Calc. c. (l.), Carb. v., Fago., *Grap.*, Hur., Mez., Na. m., Na. m. (l.), (r.), Nit. ac. (l.), Rhodo (l.), Rhu. v. (r.), Til., Verat., Verat. v. (r.); NOON, Fago.; EVENING in bed, Sul. (r.); NIGHT in bed, Merc. i. f. (r.); < night, Aur. m.; > rubbing, Zn. (l.); > scratching, Caus. (r.), Mag. c. (r.), Mag. m. (l.), Rut. (l.); ACUTE, *Mez.*; he wants to scratch the ears off, Thu.; EXTENDING to nape of neck, Rhodo. (l.); PERSISTENT, Lyc.; as from TETTER, Hur.

—LUMPS, Bar. c., Dro., Grap., Pho. ac., Sars.; hard, Cinnb. (l.), and painful to touch, Grap. (r.); sensation of, on turning head to left, *Grap.*

—MOISTURE, Aur., GRAP., Lyc., *Petrol.*, Rhu. v. (r.); and sore places, Grap.

—NUMBNESS, on pressing teeth together, Aloe.

—OPPRESSION, Thea.

—OOZING and bad smell, with red, rough, tetter-like skin in front of ear, and sticking from ear to head, Olnd. (l.).

—PAIN, Arum. d. (r.), Calc. ph., *Chel.* (r.), Castor. (l.), Glo. (l.), Hur. (r.), Kalm., Myric. (l.), Pte. (r.), Thea.; 3 A.M., Dios.; 8 A.M., Guan.; 11 A.M., Dios. (l.); 3 P.M., Pte. (r.); 5 P.M., on waking, Pte. (r.); 6 P.M., Pte. (r.); 6.30 P.M., Yuc. (r.); 10 P.M., Pte. (r.); NIGHT, Merc. i. f. (l.); on PRESSURE, Nicc. (r.); > RAISING head, Ig.; and in right side of NECK and THIGH, Kalm. (r.); in neck and cheek at night, Kalm. (r.); ACUTE, Glo., Hur., Pte. (r.); morning, Dios.; 9 A.M., Pte. (r.); DEEP-SEATED, Dios. (r.); EXTENDING in front of ear and to angle of jaw in morning, Dios. (l.); into ears, All. c.; down side of neck, Pic. ac. (r.); upward, Fl. ac. (r.); upward and diagonally towards opposite ear and right parietal bone, Cer. b. (l.); from deep in head, All. c.; drawing from neck, Apis (l.); FINE, Dios. (l.); as from pressing on a SORE, and in ear and larynx, Lyc.; WANDERING, Na. slc. (l.).

—PECKING, All. c. (r.).

—PIMPLES, Alum. (r.), Cala., Calc. (l.), Caus., Lyc. (l.), Nicc. (r.), Na. m. (l.), Pul. (l.), Saba. (r.), Sel., Sul. (r.), Sul. (l.); after SCRATCHING, Mez.; BLEEDING easily, Opu. (l.); BURNING, on touch, Canth. (r.); HARD, Grap.

(r.); INFLAMED, Sul. (r.); ITCHING, Rhu. t.; PAINFUL, Cannab. s.; SORE, Pal. (l.); on pressure, Grap. (r.), Ham.; on scratching, Mez.; on touch, Calc. p., Dro. (l.).

— PINCHING, Carb. v. (l.), Paeo. (r.); 8 A.M., Guan.

— PRESSURE, Acon., Asar. (l.), *Bell.*, Bor., Cad., Canth. (r.), Caus., Cina., Coloc. (l.), Hell., Led. (r.), Manc., Mez. (r.), Na. slfc. (l.), Nit. d. s., Plat., Rut., Stan., Thu., Verb., Vio. o.; obliging him to BITE teeth together, Crot. h.; as from a BLUNT INSTRUMENT, Cannab. s. (r.); DRAGGING, Merl.; DULL, evening on drinking rapidly, Na. m.; as from a HARD BODY, of the size of an egg, Grap.; SUDDEN, Verb. (r.).

— PULLING, K. ca., Merc. sol. (l.).

— PULSATION, All. c. (r.), *Aml. n.*, Calc. p., Caus. (l.), Glo., K. ca., Mez. (r.), Rhu. t. (l.); (AGG.) moving head, K. ca.; warmth and lying on affected side, Rhu. t. (l.); (AMEL.), cold air and walking, Rhu. t. (l.); EXTENDING into eye, Pic. ac.; HAMMERING, Cap.

— PUSTULES, Phyt.; in a circle, Cast. eq. (l.); containing serum, Sum. (l.).

— RASH, Ant. cr.; gritting, itching, Na. m.

— RAWNESS, Petrol.

— REDNESS, Acon. l., Ant. s. (r.), Nit. ac. (l.), *Petrol.*, Rhu. v., Til.

— SCABS, Grap., Pul., Sil., Stap.; exuding glutinous moisture, sore to touch, Thu. (r.); herpetic, K. iod.

— SHOOTING, 6 P.M., Erio. (r.); inward, <warmth and lying on affected side, >cold air and walking, Rhu. t. (l.).

— SORENESS, Anac., Cic. v. (l.), Cup. ar., Grap., *Petrol.*, K. ca., Nit. ac., Pso., Lyc. (r.), Mur. ac. (r.), Sul., Verat.; on TOUCH, >scratching, Rut. (l.); as from a BLOW, Cic. v., Verat.; HUMID, Petrol.; in a SPOT, Calc. p. (r.), *Grap.*; to touch, Merc. sol. (l.).

— STITCHES, Æth., *Arn.*, Aur., *Bell.*, Bro. (l.), Cannab. s. (r.), Canth., *Caus.*, Cent., *Cina*, Con., Cop., Dig., Hell., Hep., Kalm., K. ca., K. n., Mag. c., Meny., *Nitr.*, Saba., Sabi., Sars., Tab. (l.), Tarax., Verat., Verb. (r.), Vio. o., Vio. t. (l.); MORNING, Calc. c. (l.); AFTERNOON, Pho. ac.; EVENING, Sul. (l.); 1 P.M., Na. c. (r.); during REST, Sabi.; WITH stitches in ear, Am. car., Bell., Kalm. (r.); pinching in ears and with decrease of pain, Nitr. (r.); reddish swelling, Tab.; ACUTE, Verat. (l.); BITING, Bro. (l.); burning CRAWLING, Saba.; DULL, Arn.; 3 P.M., > pressure, Mag. c. (r.); EXTENDING into jaws, K. n. (l.), Lyc.; stinging of INSECTS, in a dream, Pho.; ITCHING, Vio. t. (l.); with reddish swelling, Tab.

— SWELLING, Bar. c. (l.), Benz. ac.; HARD, red, Tab. (l.); PAINFUL to touch, Cap.; REDDISH, Ant. s. (r.); with sticking, Tab.; SOFT, painless, two inches in diameter, like the swelling on a new-born child, above and behind ear, one could feel beneath the swelling a sharp edge of bone, Bar. m. (r.).

— TEARING, Aga. m., Alum., *Ambra* (l.), Am. car., Am. car. (l.), Angu., Arg., Arg. n. (l.), *Bar. c.*, Bar. c. (r.), Bell., Camph., *Canth.*, *Cap.* (l.), Chel. (r.), Colch., K. ca., K. n., Lau., Lyc. (l.), Mur. ac. (l.), Meny., *Nitr.*, Phel., Pb., Rhodo., Rhu. t. (l.), Sars., Sep., Sil. (r.), *Squ.*, Squ. (l.), Tab. (r.); 1 P.M., Na. c.; 3 P.M., Phe.; 4 P.M., Caus. (l); 9 P.M., >sitting up in bed, Alum.; <moving head, Am. car. (r.); DRAWING, Coc. c. (l.); extending towards nape, Mur. ac.; to shoulder, Ars.; EXTENDING to vertex, occiput, nape,

and shoulder, after dinner, <moving head, Am. car.; upward, K. ca.; afternoon, Sars. (l.); FINE, 1 P.M., Sil.; and below ear, in afternoon, Sil. (r.).

BEHIND EXTERNAL EAR. TENSION, Alum. (r.), Am. car., Apis (r.), Apis (l.), Asar., Caus., *Con.*, Daph., K. n., Mez. (l.), *Nitr.*, Pb., Verb.; <PRESSURE, Glo. (r.); AND beneath ears, Apis; EXTENDING from neck, Apis (l.); upward, transient, Am. car. (r.); with sticking and tearing, Nitr. (r.).

—TETTER, Am. m.; and tetter-like roughness, and below ear, Mar.; scaling off and improving, Grap.

—THUMPING, Hell. f. (l.).

—TICKLING, Bro.; >scratching, Bro. (r.).

—TUMOR, suppurating, Phyt.

—TWISTING, Am. m. (l.); extending to temples, Ced. (r.).

—TWITCHING, Am. m., Fl. ac. (l.), K. ca., Merc. sol.

—VESICLES, Calc. cau., Chin., Na. m. (l.), Pho, *Pso.* (r.), Rhu. v., Rhu. t.; filled with serum, Rhu. v.; discharging turbid serum, Rhu. t.

—WART-LIKE GROWTHS inflame and ulcerate, Calc. c.

—WEN, existing from infancy, discharged, Merc. i. r. (r.).

—BONE, pain extending towards neck, Lith. (l.); periosteal swelling, Carb. a.

—FOSSA, tearing, Carb. v. (r.).

—GLANDS swollen, Colch. (l.), Dig., *Grap.*, Hur. (r.), Nab. (r.), *Nit. ac.*, Wies.; sticking and tearing, extending through ear, at 6 P.M., >warmth of bed, Nit. ac.; tension, *Grap.*

—HAIR matted, Chel.

—MASTOID CELLS. Acute inflammation, CAP.; pain, Sap. (l.).

—MASTOID PROCESS. ACHING, Æth. (l.), Caus., Coca (r.), Con; before midnight, in warmth of bed, Coc. c. (l.); passing through to opposite side, 11 A.M., while sitting, Erg. a.; special ACTION, *Glo.*; AFFECTION, in scrofulous individuals, after abuse of mercury, Hep.; BORING, Oni., Na. slfc. (r.); BRUISED sensation, on pressure, *Cina*; extending to clavicular and pectoral regions, Coc. c. (l.); CARIES, Aur., Nit. ac.; and of ossicula, with discharge, Aur., Fl. ac.; COLDNESS, Æth. (r.); CONGESTION, burrowing, when worst, extending to clavicular region, lower back teeth, and side of occiput, Coc. c. (l.); CONSTRICTION, as by a hot band, extending from one to other across occiput, Coc. c.; CUTTING, twitching, Sil.; DRAWING, Canth. (r.), Gent. c.; in evening, Ol. an.; extending backward, towards noon, Aloe (r.); downward, Thu. (r.); >pressure, Arg.; to lower teeth, Mez. (r.); stinging, extending to left frontal protuberance, Sars. (r.); NUMB sensation, as if head were screwed together, PLAT.; PAIN, Bry. (l.), *Chel.*, Lob. s., Physo., Phyt. (l.), Pte. (r.), Verat. v. (r.); in morning, Ham.; 7 A.M., on waking, Trom. (l.); afternoon and evening, on first going out in wind, Ham.; erratic, Polyg.; extending across occiput to over right ear, Lac. ac. (l.); to shoulder, back of clavicle, at 9.30 A.M., Hydrs. (r.); to scapula, Hydrs. (l.); as if a dull nail were forced into head, Olnd.; penetrating, cramp-like, in morning, Sul.; pulsating, Hur.; afternoon, >warmth, Ir. v.; before midnight, in warmth of bed, Coc. c. (l.); suppurating, on touch, Calc. c. (l.), Coc. c. (l.); PECULIAR sensation, Ham.; PINCHING, periodical, as from pressure of a button, Thu. (r); PRESSURE, Bell., Hcy. ac., Lach., Nit. d. s. (l.); bruised, on pressure, Plat. (l.); as from a button,

Thu. (r.); drawing, Thu.; extending to opposite side, 11 A.M., while sitting, Ery. a.; outward, K. bi. (l.); tensive, Coc. c., Coc. c. (l.); extending to clavicle and lower back teeth, < lying on ear, Coc. c. (l.); towards each other, in morning, Sabi.; SENSATION, Ham.; SHOOTING, Plan.; when walking in wind, Ham.; with tearing, Canth.; outward, Calc. p.; SORENESS, Brach., Ham. (r.); extending half-way down jaw, on moving jaw, Ir. v.; STITCHES, Aga. m. (l.), Canth. (r.), Con., Euphr. (r.), Ple. (r.), Ter. (r.), Thu., Thu. (r.); acute, coarse, Cannab. s.; drawing, Bar. ac. (l.); dull, Cham. (r.); extending to anterior part of neck, morning, on rising, <motion, Na. m. (r.); intermitting, acute, Caus. (r.); SWELLING, with redness, boring pain, and feeling of obstruction in the ear, sometimes going off with a report, Sil.; painful, with inflammation, Merc.; sore to touch, CAP.; sensation of swelling, Calc. c. (l.); TEARING, Arg. (l.), Berb. (l.), Calc. cau. (r.), Canth. (r.), *Indg.* (l.), Mang. (r.), Sep. (r.), Ton. (r.); with shooting, Canth.; drawing in evening, Thu. (l.); extending into lobule of ear, Canth. (r.); upward, Rat. (l.); as with a knife, Canth. (r.); sticking, Meny.; twitching, Rhu. v. (r.); TENSION, cramp-like, Sul. (r.); dull, < after pressure, Glo. (r.); THRUSTS, Bell.; TINGLING, in afternoon, > warmth, Ir. v.; TWITCHING, Kre.

EXTERNAL EAR. ABOVE MASTOID PROCESS, pressive drawing, Chel. (l.); pain in evening, Chin. s. (r.); pulsating, Kre. (l.).

— BEHIND MASTOID PROCESS, ACHING, Acon. (r.); as if stiff at 12.40 P.M., Equ. (l.); PAIN, acute, extending outward in evening, Ost.; extending backward and upward, Cer. b. (l.); to upper part of orbit, < after sleep, Na. hy. (r.); tingling, All. c. (l.); STITCHES, Vio. (r.); 7 to 8 P.M., Lyc.

— BELOW MASTOID PROCESS, BRUISED sensation, RUT.; DRAWING, in evening, Thu. (r.); >pressure, Dig.; cramp-like, Sul. (r.); NODE, burning, pulsating, Eupi. (l.); PAIN, 9 A.M., on pressure, Glo. (l.); REDNESS, afternoon, Fago.; STITCHES, like pinching, *Cina;* TENSION, in evening, Thu. (r.); at 4 P.M., Bry.

— MUSCLES OF MASTOID PROCESS, CRAMP, drawing, Mang. (l.); DRAWING, Lau.; sprained, extending to clavicle, Petrol.; PAIN, Bell.; TEARING, extending to clavicle, Petrol.; TENSION, cramp-like, during day, Sul.

— PERIOSTEUM OF MASTOID PROCESS, stitches, after 7 P.M., Carb. a. (r.); swelling, Guare.; after 7 P.M., Carb. a. (r.).

— PETROUS PORTION OF MASTOID PROCESS, swollen, red, and painful, Cap.

— POSTERIORLY IN MASTOID PROCESS, pressive, cutting, and suppurating pain, on touch, Mur. ac.

— REGION OF MASTOID PROCESS, tumors or boils, Con.

— ROOT OF MASTOID PROCESS, dull stitches, >touch, Sars. (r.).

— MEATUS, on level with, lancinations, 10.30 P.M., Ir. fœ. (r.).

— NECK, on. PAIN, Cham. (l.), Rap.; pulsative, Angu.; PIMPLES, easily bleeding, Opu. (l.); tearing PRESSURE, Iod. (r.); twitching STITCHES, evening, Stap.; TENSION, Apis.

— SKIN, swelling, Rhu. v.; tension, Con.; spasmodic, extending obliquely into nape, Lyc.

— SOFT PARTS, pressure, Led. (r.).

BENEATH EXTERNAL EAR. ACHING, Colch., Opu.; BRUISED sensation on touch, Zn. (r.); BURNING, 10 A.M., Rut. (l.); COLDNESS, Cer. s.; CONGESTION, burrowing, Coc. c. (l.); CONTRACTION, cramp-like, extending to ramus of lower jaw, Dul. (l.); CRACKING, Chel. (l.); CRAWLING, Verat.; DARTING, Xan. (r.); DRAWING, Sul. ac. (r.); outward, cramp-like, Olnd.; ERUPTION discharging water, Pte. (r.); dry scaly, Mar. (r.); FURUNCLE, with tension in joint of jaw when chewing, Calc. ostr.; ITCHING, Ars. (r.), Caus., Verat.; > scratching, Mag. c. (r.), Ol. an. (r.); biting, Verat.; LANCINATION, Tarent. (l.); PAIN, Acon. (r.), Cap., Chel. (r.), Opu.; when swallowing, Na. hy. (r.); PAPULES, Ars. (r.), Euph. a. (r.); itching, Mag. c.; PRESSURE, Arum m., Asar. (l.); inward, Sep.; PULSATION, Sang.; irregular, Sang.; RENTS, Olnd., a single rent, Chel.; afternoon, > after pressing, Sul.; SHOOTING, 6 P.M., Erio. (r.); has to cry out, Bar. c. (r.); SMARTING, burning, during menses, Mag. c. (l.); SORENESS, Colch., Na. hy. (r.); STITCHES, Apis (l.), Crot. t., Mag. s. (l.); deep, each alternately, Vio. o. (< l.); pressive, Coc. c. (l.); SUPPURATION, painful, Na. hy. (r.); SWELLING, Berb. (r.); extending up head, Na. hy. (r.); hard, painful, < pressure, All. c. (l.); great swelling, heat, redness, and a lump, with sharp pain, restless at night, with improvement of hearing, Sam. (r.); TEARING, Ol. an. (r.), Pho. (r.); while sitting, > rubbing, Pho.; extending sometimes to vertex, occiput, nape, and shoulder, after dinner, < moving head, Am. car; fine, and below, 1 P.M., Sil. (r.); transient at 4 P.M., Caus. (l.); and behind upward, violent, towards the helix, after five minutes it goes slowly towards the left shoulder, then into nape of neck, and finally into the occiput, Am. car. (l.); TENSION, Apis, Apis (r.), Spig. (l.); and behind, Apis; TETTERS and tetter-like roughness, and behind, Mar.; THREAD tied around, sensation of, Rum. c.; THUMPING, Hell. f. (l.); TICKLING, Sang.; VESICLES, Pte. (r.); BONES, pressing construction, Zn. (r.); CAROTIDS, pulsation, Verat. (l.); FOSSA, boring and bruised pain on pressure, Caus. (r.); tearing pressure, Iod. (r.); GLANDS, inflammation, Sars. (r.); soreness, Pte. (r.); extending into middle of posterior cervical region, < evening, Cinnb. (r.); on touch, BAR. C. (r.); swelling, BAR. C. (r.), Pte. (r.), Sars.; hard, Am. car.; near JAW, burrowing, Lach. (l.); stitch, Bar. c. (r.); tearing and pain on touch, Colch. (r.).

MIDDLE EAR.

Middle Ear. ABSCESS would form, sensation as if, with heat and a stitch, Bry.

ACHING, Canc. f. (r.), (Chin.), Saba.; in day, Stach. (l.); on swallowing, Dro., Fago. (l.).

AIR penetrates on blowing nose, Pul., Sul.; on drawing jaw to other side, Sarr. (r.); cold air rushes in during eructations, Caus.; passing out of cold, as when laughing, Mil. (l.).

BORING, WITH burning in throat, Ol. an.; obstructed feeling, sometimes going off with a report, and swelling and redness of the mastoid process, Sil.; outward, Euphr. (r.).

BUBBLING, Con., Dul. (l.), Euphr., K. ca., K. n., Petrol., Rhe., Sil. (r.); on stooping, Grap.; as of air, Hur. (l.), Lyc., Na. m. (l.); deep-seated, Lim.

BURNING, Angu.; as from a coal, Tep. (r.); painful, on swallowing, Hæm.

BURSTING SENSATION, when eating, or swallowing saliva, Pso.; as of bubbles, Carb. v., Grap., Na. m. (r.); of a membrane, then buzzing, Gam.; painful, from the throat towards the ear, All. c.

CATARRH, purulent, Vesp.

CLOSED SENSATION, on blowing nose, Mar. (r.); from within, when swallowing, Ars.

CONTRACTION, in evening, Spig.

CRAMP, in evening, and in external ear, Ran. b. (l.); like ear-ache and compression, in evening, Thu.

CRACKLING, < chewing, Alum.

CRUSHING, when swallowing, *Calc. c.*

CUTTING, K. iod.

DARTING, acute, Ton. (r.).

DISTENSION, painful sensation of, Til.; as if it would burst, or as if something struck the drum, Sil.

DRAWING, Angu. (r.), Pho. ac. (r.); during ERUCTATIONS, Sul. (l.); on SWALLOWING, Fer. ma.; as if ears would burst, Hell.; EXTENDING into Eustachian tube, after dinner, Ant. c. (r.); SHOOTING drawing together, on swallowing, Dro.; SPASMODIC, at night on swallowing, Alum.

DRYNESS, Colch.

FLUTTERING, Aga. m. (r.); as if opening and shutting, in right, then left, then both at same time, > putting in finger, Ir. fœ.

FORCED in, feeling as if something were, Lyc.

FULNESS, in evening, Na. p.; after stitches leave it, Iod.; from swallowing, Arum d.; with heat, at 1 P.M., Com.

HEAT, Arum d. (l., then r.), Bry., *Calc. c.;* MORNING in bed, Cocc. (r.); from

SWALLOWING, Arun.; WITH heat in external ear, Calc. p., Casc.; morning in bed, (Cocc.); fulness, 1 P.M., Com.; inflammation of external ear, redness and swelling, Pul.; stitch and sensation as if an abscess would form, Bry.; like HOT BLOOD, Calc. c.

INFLAMMATION, Acon., Chin., Cocc., K. ca., MERC. SOL.; WITH inflammation of external ear, Bell., Calc. ostr., Merc., and discharge of pus, Bell. (r.), and redness and swelling, Pul.; discharge, Carb. v.; stinging and tearing, Merc.

ITCHING, Dios., Spira.; EVENING, Elap.; 10.30 P.M., Dios. (r.), (l.); when SWALLOWING, *Sil.;* WITH crawling, when swallowing, and itching in throat, K. ca., Sum.; desire to swallow, Nx. v.; inclination to lessen it by swallowing, Carb. v.

OBSTRUCTED FEELING, sometimes going off with a report, with boring pain in ear, and redness and swelling of mastoid process, Sil.

OPENING through which air could penetrate, sensation of, on opening and closing mouth, with thrust-like stitches from right side of fauces suddenly extending into ear, Thu. (r.).

OPENING AND SHUTTING SENSATION, 6.45 P.M., Ir. fœ.

PAIN, Apis (r.), Arum d. (r.), Coca (r.), Ig., Physo. (r.); AFTERNOON and EVENING, Sul. (r.); AFTERNOON and during ERUCTATIONS, Tarent. (r.); on SWALLOWING, with ulcers, Bov. (r.); swallowing saliva, Pb.; on TOUCHING external ear, Tab.; WITH contracted feeling in throat, Hæm.; warmth, coldness of right concha and burning heat of left, which extended beyond the temple, in evening, Na. n.; EXTENDING from throat, Lith.; a painful spot in throat, Lob. i.; to ear, when swallowing, Lach. (l.).

PINCHING, Am. car., *Carl.* (l.); with acute thrusts, *Bell.;* extending towards drum, Dul. (l.); into pharynx, Carb. a. (l.).

PRESSURE, Calc. cau., Physo. (r.); on SNEEZING or SWALLOWING, *Sul.;* as from a BLOW, on stooping, Cham.; as if it would BURST, Rhu. t.; with stitches, Lyc.; INTERMITTENT, Arn.; INWARD, Nit. ac. (l.), Tarax. (l.); OUTWARD, Na. slfc. m.; on loud reading, Coca; on swallowing, Nx. v.; STICKING, Calc. cau. (l.); as if every thing would be pressed TOGETHER, almost cramp-like, Dro. (r.); like TWINGING, Grap.

PULSATION, Nit. ac., Sil. (r.).

RATTLING, Aga. m. (r.).

RELAXED SENSATION, with hard hearing, <violent swallowing, Rhe. (r.).

ROLLING, with twitching, Aga. m. (r.).

SHOCKS, or twinging, waking from sleep, Merc. d. (l.).

SHOOTING, Bell.; during ERUCTATIONS, Bell.; shooting from ear, <SWALLOWING, Con.; WITH pain in throat, K. ca.; EXTENDING into palate, K. bi.; STINGING, Tarent.

SOMETHING before, Calc. a.; sensation of, on blowing nose hard, >swallowing, Calc. c.

SORENESS, Cup. ar. (r.); in afternoon, Dios. (l.).

SORENESS, dull, in *r.* internal ear, Cup. ar., *both ears are internally sore and denuded,* the right worse, *Merc. sol.*

SORENESS of *l.* internal ear, in P.M., Dios.; both ears sore to touch internally, 8 P.M., Dios.

STABBING, Physo.

STITCHES, Bry., Canc. f. (r.), *Ran. b.* (r.), Rhodo., Tanac., Thu. (r.); when BLOWING NOSE, Hep., Lyc.; when DRINKING, Com.; during SWALLOWING, Na. m. (r.); empty swallowing, Thu.; WITH stitches in larynx when swallowing, Mang.; inflammation and tearing, Merc.; warmth in ear, and sensation as if an abscess would form, Bry.; COLD, Aga. m. (l.); CUTTING, extending to brain, Arg. (l.); DRAWING, Aloe (l., then r.); EXTENDING outward, Sep.; through external ear, Thu. (r.); inward, Arn.; FINE, Dro. (l); ITCHING, *Pul.;* LIGHTNING-LIKE, in evening, Thu.; PICKING, with burning of external ear, Cle.; SCRAPING, Mang.; TEARING, alternating with same in other parts of head, Berb.; THRUST-LIKE, sudden, coming from right side of fauces, with sensation, on opening and closing mouth, as if there were an opening in ear, which air could penetrate, Thu. (r.).

STOPPED SENSATION, Glo.; morning on rising, >blowing nose, Stan. (l.).

SWELLING, with discharge, Cist.; inflammation of middle and external ear, redness and heat, Pul.

TEARING, *Ars.*, Berb., Caus., Chel., Pho. ac.; AFTERNOON, when sitting, Indg.; EVENING, Merl.; WITH tearing in external ear, Mag. m., Rat., and in cartilage, K. ca.; stinging and inflammation, Merc.; tearing in right side of head, on raising head after stooping, Ant. t. (r.); ASUNDER, Con.; EXTENDING downward, and in external ear, BELL.; FINE, Cyc. (l.); SUPPURATIVE, <boring in, Anac.

THRUSTS, with pinching, Bell.

TICKLING, Pul.; extending into Eustachian tube, Na. p.; crawling, Mang.

TWINGING, Dro. (l.); and in external ear, Asar.; or shocks, waking from sleep, Merc. d. (l.).

TWITCHING, Am. car., Lyc.; on BLOWING nose, and on SNEEZING, Act.; AND in external ear, Dig.; with rolling, Aga. m. (r.); EXTENDING to shoulder, Cannab. s. (r.); TEARING, Angu. (r.).

ULCERATED SENSATION, on swallowing, *Sul.*

Eustachian Tube. ACHING before dinner, <turning head to right and on swallowing, >warm soup, Coc. c. (r.); AFFECTION, Lyc.; AIR CATCHES ITSELF, as in a sac, on taking a pinch of snuff, and on eructation of wind, Tel. (l.); PRICKLING BURNING, <swallowing, Acon.; CATARRH, Calc., Con., Gel., Grap., Iod.; with stitch-like pain and tearing in ear, and redness and swelling in meatus, Pul.; CUTTING on chewing, Arg.; DRAWING, Pul. n. (r.); after dinner, Ant. c. (r.); DRYNESS, Stram. (r.); FLAPPING, as if air were forced through, on every eructation, Grap.; GURGLING as from air, Caus., Grap.; INFLAMMATION, Ery. a.; IRRITATION, Phyt.; ITCHING, Caus., Nx. v.; sensation of MUCUS in, Cot. (l.); OBSTRUCTION, Hydrs., Phyt. (l.); PAIN, Cot. (l.), Lach., Ox. ac.; acute, to submaxillary gland, 8 P.M., Fago. (r.); as from a rough body (on change of weather) before the setting in of wind or rain, Nx. m.; STITCHES, Aga. m. (r.), Carl., Menth. pu. (l.); extending within ear, on swallowing, Sal. ac. (r.); itching, extending to ear, <boring with finger, Coloc.; stinging, Tarent.; twitching to drum, >boring with finger, *Carl.;* STUFFING sensation, Acon. (r.); running of WAX, Crot. h., Lach.; in OPENING IN EAR, pain now and then, in afternoon, Ox. ac. (<r.); at PHA-

RYNGEAL ORIFICE, itching crawling to drum, Arg. (r., then l.); tickling itching, extending into tympanum, alternating with ringing in left ear, Aga. m. (r.); pain in afternoon, Ox. ac. (<r.); in region of, chilling or burning twitching, then above it a sort of gnawing, both painless, All. c. (r.); VOLUPTUOUS, extending through inner ear to mouth, Coc. c.

MUSCLE, tensor tympani, jumping, Aga. m. (r.).

OSSICULA, caries, and of mastoid process, with discharge, Aur., Fl. ac.; caries with inflammation, Lyc., Sil., Sul.; destruction and discharge of, Asaf., Aur., Hep., Nit. ac., Sil.

INTERNAL EAR.

Internal Ear. HEARING ACUTE (includes SENSITIVE, IRRITATED), Acon., Aga. m., Alco., Alum., Ambra, Am. car., Am. mur., Anac., Angu., Ant. cr., Apis, Arn., Ars., Ars. hydr., Asaf., Asar., Aur., Bar. c., BELL., Bor., Bry., Cact., Cala., Calc. c., Calc. ostr., Camph., Cannab. i., Canth., Cap., Carb. a., Carb. v., Caus., Ced., Cham., Chel., Cic. v., Cic. v. (l.), Cina, Cinch., Cocc., Coff., Colch., Coloc., Con., Cup., Dig., Fl. ac., Gam., Grap., Hell., Hep., Hydrphb., Hyos., Hype., Ign., Iod., Ip., K. ca., K. hydro., Lach., Lau., Lyc., Mand., Mag. c., Mag. m., Mar. (r.), Merc., Mos, Mur. ac., Narcot., Na. c., Na. slfc., Nit. ac., Nit. ox., Nx. m., Nx. v., Olnd., Ol. an., Op., Ox. ac., Petrol., Pb., Pho., Pho. ac., Physo., Phyt., Phyt. (<r.), Plan., *Plat.*, Pul., Saba., Sabi., Sang., Sars., Sec. c., Sel., Seneg., *Sep.*, Scu., SIL., Spig., Squ., Stan., Stap., Stram., Stry., Sul., Tab., *Ther.*, Thu., Val., Verat., Vio. o., Zn., Zing.; MORNING, Fl. ac., Rhodo.; EVENING, Calc. a., Coca, Grap., Rhodo.; on falling asleep, Calc. a.; NIGHT, Bry., Carb. v., Atro.; after waving in HEAD and cracking in ear, Grap.; when LYING, Grap.; during MENSES, Hype.; hears every thing in SLUMBER, Alumn., Ars., Grat., Na. m., Op., Sul.; in partial slumber, Euphr.; and retains a clear consciousness of his condition in a sort of slumber, Mor.; on waking, Carb. v., Pul.; on WALKING, Lyc.; >open AIR, Tab.

—SENSITIVE, WITH impaired HEARING, Am. car., Arn., Lyc., Merc., Plat., Sul., and anxiety, Pul., Sil.; crawling in ears, Lach.; roaring, Acon.; ringing of slightest tone, wakes with rush of blood to head, hair standing on end, anxiousness and shuddering, formication from the slightest motion in bed, Carb. v.; illusions, Cap., Cup., Ol. an.; great affection of MIND, Zn.; desire to be alone, Con., and at rest, Bell.; anger, Ip.; anxiety, Aur., Cap., Pul., Sil.; hatred of company, Bell, Pho.; bitter complaints, Ign.; irritability, Bell., Calc., Con., Hype., K. ca., Nx. v., Pho.; whining and crying, Crot. h., Lach.; HEADACHE, *Anac.*, Apis, Bar. c., Bell., Calc. ostr., Con., Ig., Iod., Merc., Nit. ac., *Pho. ac.*, *Spig.*, Stan.; contracting, Acon.; shooting, Cic. v.; tearing, Lach., Spig.; dizziness, Ther.; bursting, Spig.; irritability of head, Bar. c., Calc. ostr., Nit. ac., Pho.; stunning, through brain, Stan.; rush of blood to head, Pul.; desire to shut EYES, Con.; shunning of light, Con.; enlarged pupils, Ign.; TOOTHACHE, Calc. ostr., Ther.; NAUSEA, Nx. v., Sul., Ther.; crackling noise in ABDOMEN, Merc.; DIARRHŒA, Cocc., Nit. ac., Nx. v.; LEUCORRHŒA, Carb. v.; COUGH, Arn., Pho. ac.; CONVULSIONS, Nx v.; tetanic spasms, Cic. v.; tetanus, Castor., Nx. v.; AGGRAVATION OF PAINS, Arn., Ign., Iod.; RESERVED MANNER, Con.; acuteness of all SENSES, Coff.; SORENESS all over, Coloc.,

Mag. m., Nx. v.; STARTING, Narcot., Na. car., Saba., K. hydro., Con., Sil., *Mag.* car.; with shock through the whole body, Ther.; frightened, Ant. cr., Calc. ostr., Calc., Cannab. s., Card. b., Con., Hype., NA. CAR., Saba.; when the door is opened, Mos.; starting up from the sofa with his whole body, Carb. v.; with shuddering, Carb. v.; out of sleep, Angu., Apis; whistling sound through every LIMB, Grap.; with TORPOR, Op.; with COLD STAGE, Arn., CAP.; HORRIPILATION, Sang.; HEAT, Con.

SENSITIVE HEARING, ALTERNATING with dull, Anac.

— BELLS, causing stinging in ears, Pho. ac.; church-bell is doleful, and moves to tears, Ant. cr.

— CARRIAGES, rattling of, Nit. ac.; during deafness for voice, Chen. a.

— COCKS CROWING, causing headache, nausea, and vertigo, Ther.

— CLOCKS striking and COCKS crowing at a distance keep her awake, Lyc.

— HAMMER-STROKE in a neighboring smithy, every, Sang.

— LAUGHING, Mang

— MINUTE SOUNDS, Phyt.

— MUSIC (including MUSIC UNBEARABLE), ACON., Ambra, Am. car., Anac., Ant. cr., Bry., *Calc. ostr.*, Carb. a., Carb. v., Caus., *Cham.*, Coca, Coff., Croc., *Dig.*, Grap., Ign., K. ca., Kre., *Lyc.*, Mang., Merl., NA. C., Na. slfc., NX. V., Ol. an., *Pho.*, PHO. AC., *Pul.*, Rhodo., Saba., Sabi., Sars., Seneg, SEP., Sil., Spo., Stan., Stap., Stict., Sul., Tab., Ther., *Thu.*, *Vio. o.*, Zn.; even in the DEAF, Sul.; tunes which he had formerly LIKED, Seneg.; AVERSION to music, Acon., Carb. v., Cham., Na. c., Nx. v., Pho., Pho. ac., Sabi., Sep., and every thing, Merc.; especially to the violin, Vio. o.; with peevishness and irritability, Caus.; seeks solitude, darkness, and silence, Nx. v.; cannot bear it in the head, Pho.; nervousness so great that it goes through bone and marrow, Sabi.; when merry, Croc.; CAUSING dejection, and increase of fearful presentiments, Dig.; irritability, Mang., Na. slfc., Sabi.; melancholy, Acon., Na. slfc.; sadness, Acon., Dig.; sorrow, Na. slfc.; inclination to weep, Ant. cr., Grap., Kre., Nx. v., Thu., < in evening, Grap., and other uncommon emotions, Kre.; inclination to weep from even lively music, Na. slfc.; a single note sets him to singing, Croc.; causes headache, Ambra, Pho, Ther.; <headache, Ign.; causes congestion of blood to head, Ambra; vertigo, Ther.; weakness of head, Pho.; pressure on occiput, Sabi.; stinging pains in ears, Pho. ac., Tab., and drawing sensation in cheeks and teeth, Pho. ac.; trembling sensation in ears, Saba.; aggravation of toothache, Calc. ostr.; a single note causes nose-bleed, Hep.; music causes nausea, Sul., Ther.; cough, Ambra, Calc. ostr., Cham., K. ca., Kre., Pho. ac.; a single note causes cough, Stan.; music causes painful anxiety in chest, Na. c.; aggravation of symptoms of disturbed circulation, Acon., Calc. c., Dig., Lyc., Nat., Nx. v., Pho. ac., Sep., Stap., Thu., Vio. o.; palpitation, Stap.; orgasms, Ambra; drawing in upper limbs, Merl.; complaints, Calc. ostr., K. ca., Vio. o.; uncommon and agreeable sensations, Ig.; trembling, Na. c., Na. slfc., in evening, Sabi.; trembling and weariness, must lie down, Na. c.; weakness, Na. slfc., Nx. v., Sabi., Sep.; a single note causes exhaustion, Nx. v.; made sick by dull hearing of music, Sul.; music makes her drowsy, and shuts the eyes, a clairvoyant dream, Stan.; sweat, >open air, Sabi.; DANCING

MUSIC causes dreams, Mag. s.; MERRY TUNE is softly repeated when vexed, Croc.; the ORGAN fatigues, Lyc.; PIANO PLAYING is intolerable, Na. car.. Na. slfc., *Pho.*, Sep.; complaints from, Anac., Calc. ostr., K. ca., Merl., Na. c., Na. slfc., Nit. ac., Sep., Zn.; causes weakness of head, Pho.; nausea, Sul.; fatigue, Sep., with painful anxiety in chest, Na. c.; trembling and convulsions, Na. slfc.; SAD music mollifies ill-humor, Mang.; makes him lively, the most lively does not exhilarate, Mang.; sensitive to SINGING, Am. car., Pho. ac., Sars., Stan., Spo.; aversion to hearing singing, Lyc.; singing causes exhaustion, Nx. v.; in a church, palpitation, Carb. a.; can play only the SOFTEST tones, Coff.; abhorrence of the VIOLIN, Vio. o.

—NOISES (including NOISES UNBEARABLE), ACON., Alum , Al. p. s., Am. car., Anac., *Angu.*, Ant. cr., Apis, *Arn.*, Ars., Aur., Bar. c., *Bell.*, *Bor.*, Bry., Cala., *Calc. c.*, *Calc. ostr.*, *Camph.*, Cannab. i., Cap., Carb. a., Carb. v., Card. m., Caus., *Cham.*, Chel., Cic. v., Cina, Cinch., *Cocc.*, COFF., Coloc., Con., Crot. h., CUP., Dig., Fl. ac., Gam., Grap., Hell., Hype., Hur., *Ign.*, Iod., Ip., K. ca., K. hydr., Kre., LACH., LYC., *Mag. c.*, Mag. m., Manc., Mang., Merc., Mos., Mur. ac., *Narcot.*, *Na. c.*, Na. m., *Na. slfc.*, Nit. ac., Nx. m., Nx. v., Olnd., Ol. an., Op., Ox. ac., Petrol., PHO., PHO. AC., *Plat.*, Pb., Pte., *Pul.*, Saba., Sang., Sel., *Sep.*, Scu., Sil., *Spig.*, Stan., Stram., SUL., Tab., THER., Xan., *Zn.;* WITH hard hearing, Merc., and dreamy, dull state of mind, Zn.; roaring, Acon.; anger after contradiction, Cocc., anger and rage, Ip.; anxiety, Aur., Cap., Caus., Na. c., Pul., Sil.; full of care, Aur., Bar. c.; difficult comprehension, Cap.; cross, all things disagree, Pho.; displeased with every thing, nothing is right, Ars.; fretfulness, Ars.; ill-humor, Bell., Pho.; illusions, at night, Carb. v.; irritation and faint-heartedness, Cinch.; unreasonable lamentations, bitter complaints, Ign.; reserved mania, Con.; men are offensive, Pho., melancholy, a little noise startles much, Stram.; morose and peevish, K. ca.; peevish, irritable, Pte.; stupefaction, Lach.; suspicious, as if his life were conspired against, Al. p. s.; vexation, Rhu. t., and anger, Mang.; weeping, Æth., Kre.; CAUSING heat, *Bry.*, Caus., *Coff.*, Sep.; distress, with otitis, Merc.; tingling, Lach.; every noise, as FILING, SCRATCHING, SCRAPING with the feet, causes shuddering, which penetrates the teeth and causes trembling, Rhodi., Sul.; LOUD, Calc. ostr., Cap., Cup., Iod., Ol. an., Pho., Sil., Spig , Tab., Ther.; anxiety when among the noise of MANY PEOPLE, Petrol.; every sound or SHRILL noise penetrates the whole body, especially the teeth, causes vertigo, which produces nausea, Ther.; SLIGHTEST noise, Angu., Ant. cr., Ars. iod., Coloc., Calc. ostr., Cannab. s., Carb. v., Card. b., Cic. v., Con., Ip., Narcot., N. c., Nx. v., Op., Pho. ac., Plat., Saba., Sel., Tab., Ther.; in sleep, Ars. hydr., Cala., Op., Pho., Sel.; causing crying and weeping, Lach.; full of fears with every noise in the STREET, Caus.; SUDDEN, Phyt., Sang.

—ORGAN. See under MUSIC.

—PAINFUL, Cocc., Con., Lyc., Sang., Seneg., Sil., Spig.

—PAPER, folding of, Cala.

—PIANO. See under MUSIC.

—READING, loud, Verben.

SENSITIVE HEARING. SHARP SOUNDS, Cop.

—SINGING. See under MUSIC.

—STEP, every, Coff.; cannot bear to hear walking in the room, with extreme moroseness and nausea, Sang.

—SUDDEN, with looseness of bowels, Bell., Bor.

—TALKING OF OTHERS, Ambra, Am. car., Arn., Ars., Aur., Bar. c., Calc. ostr., Carb. v., Cinch., Cocc., Colch., Con., Ign., Iod., K. ca., Mag. mur., Mang., Mar., Nx. v., Pho., Pho. ac., Rhu. t., Sil., Spig., Stap., Sul., Ther., Ter. m. m., Verat., Verb., Zn., Zn. o.; CAUSING excitement and irritability, Am. car., and trembling through whole body, Ambra, Calc. c.; sensation as if losing consciousness, K. ca.; dulness, Stap.; fright and shooting in head, Cic. v.; confusion of ideas from the increased headache, Aur.; increased headache, Aur., Cocc., Iod., Pho. ac.; ailing in one half of head, Ign.; ailing in forehead, Sil.; rush of blood to head, Coff.; shaking of brain, Con.; sore feeling in brain, Cinch.; vertigo, Cham.; dull stitches in ear, Mang.; heat in face, Sep.; complaints, Sil.; increase of pains, Arn., Ars., Mag. m., Sul., Ther., Zn. o.; weeping when spoken to, Stap.; AVERSION to, Iod., Pho. ac., Zn.; at a DISTANCE, making him nervous, and causing headache, Mur. ac.; LOUD, Carb. v., Coff. t., Pho., Pte.; causing pain as if the head would burst, Iod.; dizzy headache as if bursting, loud, strong, causing drowsiness in head, Spig.; of MEN, causing headache, Bar. c.; her OWN, Op.

—VIOLIN. See under MUSIC.

—WALKING. See STEP.

—WATER poured out or running, if he hears, or if he sees it, he becomes very irritable or nervous, it causes desire for stool, and other ailments, Hydroph.; if the hydrant runs in his room his nervous headache becomes unbearable, Hydrophb.

—WHIP, crack of, (Sul.).

ALTERED, during COITION, Grap.; when LYING, Ant. cr., Agn., Aur., Con., Cro., Grap., Hep., Mag. c., Merc., Na. m., Pho., Pho. ac., Plat., Pul., Rhodo., Rhu. t., Sul., Thu., Val.; lying on affected side, Am. car., Bar. c., Euphr., Sep., Sil.; lying on back, Na. c.; when MOVING HEAD, Na. c., Pul., Stap.; resting head on table, Fer.; when scratching head, Am. car.; at REST, Stap.; RISING and SITTING up, Euphm., Grap., Na. c., Sep., Verat.; when SNEEZING, Bar. ac., Euphm.; on STANDING, Am. m, Ars., Bell., Con., Na. m., Sul.; during painful, bloody STOOLS, K. clc.; when TALKING aloud, Mar., Pho., Spig; on WALKING, Aga. m., Bar. c., Bell., Benz. ac., Carb. a., Chel., Mang., Nicc., Rhu. t., Spig.; WHISTLING, Rhodo.; > SITTING up, Am. car.; WITH a shock as of a cannon, Bad.; vertigo, Acon., Bell., Carb. v., Sang.; with vibration in head, Sil.; with ringing in head, Ars.; with flow of blood to head, roaring and humming, Sang., Sul.; headache, Glo., Plat.; in temples, Cinch.; with shaking of head, K. ca; constipation, Alum.; pain in limbs, Ars.; cold feet, Thu.

ANOTHER, as with ears of, Pso.

CONFUSED, Alco., *Carb. a.*, Equ. (l.), (< l.); does not know from what direction sounds come, Carb. a.

DISTANT, sounds seem, Cham., Eth., Sol. n.; on turning upper part of body from right to left, Eupi.; voices seem, Coca; on waking, Nit. ox.; from unconsciousness, Nit. ox.; his own voice, Cannab. i.

ECHOING. See REVERBERATING.

ILLUSIONS, Abs., Alco., Atro., Carbn. ox., Carbn. s., Conin., Elap., Eup. pur., Hyos., K. br., Stram., Thea.; during sleep, Acon. l.; with ringing, Val.

IMPAIRED (includes DIFFICULT, HARD, DULL, etc.), Acon., Æth. (< l.), Aga. m., *Agn.*, Alco., *All. c.*, Alum., AMBRA, AM. CAR., *Am. m.*, ANAC., Anac., (l.), *Angu.*, *Ant. cr.*, Apis, Agr., Arg. n., *Arg. n.* (l.), *Arn.*, ARS., *Asaf.*, Asar. (< r.), Ast., Ast. (< r.), AUR., Aur. m., Aur. s., *Bap.*, *Bar. c.*, BELL, *Bor.* (l.), *Bov.*, *Bry.*, Bry. (l.), *Cact.*, Calc. c., CALC. OSTR., Calc. ostr. followed by Lyc., Calc. p., Cannab. i., Canth., CAP., *Carb. a.*, Carb. v., *Carbn. o.*, Carbn. s., CAUS., *Cham.*, *Chel.*, Chel. (l.), Chin., Chin. s., Cle., Cic. v., Cic. v. (l.), CINCH., Cist., *Cocc.*, Cocc. (r.), Coc. c., Coc. c. (l.), Coff., Colch., Coloc, Com., CON., Conin., Cori. r., *Croc.*, Cup. ac., CYC., Cyc. (r.), Der., Dig., DRO., *Dul.*, Euphrb., Eth., Fer., Gad., Gam., *Gel.*, Glo.. GRAP., Guara., Hal., Hep., *Hydrs.*, *Hcy. ac.*, HYOS., Ib., Ib. (r.), Ig., Ip., IOD., Iodf., Jat., Kalm, K. CA., K. br., K. iod., Kre., *Lach.*, LACHN., LAU, LED., LYC., MAG. C., MAG. M., Mag. m. (l.), *Mang.*, *Meny.*, *Merc.*, Merc. i. r., Merc. sol., Merc. sol. (r.), *Mez.*, *Mos.*, *Mur. ac.*, Na. ar., NA. C., Na. c. (l.), *Na. m.*, *Nicc.*, Nico., Nitr., NIT. AC., Nit. ac. followed by Petrol., Nx. m., Nx. v., OLND., OP., Op. (l.), Par., PETROL, Petrol. followed by Nit. ac., PHO. AC., PHO., Physo. (r.), Phyt., Plat., PB., PUL., PUL. N., Ran. b., *Rhe.*, *Rhodo.*, RHU. T., RUT., *Saba.*, Sabi. *Sal. ac.*, *Sang.*, Sars., SEC. C, Sel., SEP., SIL., *Spig.*, *Spo.*, Squ., Stan., *Stap.*, STRAM., SUL., Sul. (r., then l.), *Sul. ac.*, Tab., Tarax., Tarent., *Tel.*, Tep., Ther., Thu., Thu. (r.), Val., VERAT., *Verb.*, Vio. o., Wies., Zn.; MORNING, Calc. c., Cle., Gam., Merc. i. r.; FORENOON, Asaf.; 7.30 A.M., Cle.; 11 A.M., Mag. c. (r.); till 8 P.M., Physo.; AFTERNOON, Elap., Sil.; EVENING, K. ca.; 4.30 P.M., Mag. c. (l.); 9 P.M., Physo. (r.); NIGHT, Ced.; in bruised EAR, Jac. (l.), Lach. (l.); after otorrhœa, Bor.; after pain, Nit. ac.; after burning and stinging, Cap.; after BLOWING NOSE, Con.; after taking COLD, Bry., Pul.; after cutting hand, Bell., Pul.; after a CONCUSSION, fall, firing of gun or cannon, Arn.; after DINNER, Sul. (l.); while EATING, Sul.; on mental EXERTION, Con.; after MEASLES, Merc., Pul.; checked measles, Pul., Merc.; during MENSES, Mag. m.; in OLD PEOPLE, Petrol., Verat.; after abuse of QUININE, checked intermittents, Calc. ostr., Sec. c.; when READING aloud, Coca, Verben.; in RHEUMATIC or GOUTY DIATHESIS, Petrol., Rhodo.; after SCARLET FEVER, Lyc., Nit. ac., Sul.; in SCROFULOUS SUBJECTS, Calc. i., Lyc.; after SINGING, Apoc. c. (l.), Ars., Pho.; after checked foot SWEAT, a warm sand-bath for the feet and Baryt., Sil., or Sec. c. internally; after TYPHOID FEVER, Arn., Pho.; after WAKING, Sep., Zn.; (AGG.) eating, Sil.; in house, Mag. c. (r.); noise, Ol. an., Plat., Tab.; stillness, Ant. cr.; change of weather, Mang.; (AMEL.), 10 A.M., after being out, Merc. i. r.; after dinner, Gam.; driving in a carriage, Grap.; after a crash in ear, Grap., Mur. ac., Sil.; for a moment only, Mang.; after noises, Arn., Mur. ac.; after rushing, Lach., and roaring, galvanism, Lach.; after spattering, Spig.; when snapping is felt, Tarent.; on blowing nose, Meny.; on swallowing, Merc.; after getting warm from walking, Merc. i. r.; dashing on cold water, Glo.

— WITH BUZZING in EAR, Bov., Petrol., Mag. c.; humming, Cact., *Dro.*, Iod.,

Nicc., Sec. c.; and whizzing, Lyc., Merc.; CATARRHAL AFFECTION, Calc. ostr., Con., Gel., Fer., Grap., Iod., Pul.; DIGGING, Jac.; DISCHARGE, Am. m., Asaf., Calc. ostr., Carb. v., Caus., Elap., Lyc., Merc., Sil., Tel.; of pus, Asaf., Bor., Pul., at times, Sul., fetid, Asaf., Aur., Bov.; DRAWING PAIN, Lach.; sensation as if DRUM were relaxed, Rhe.; DRYNESS, GRAP., *Mur. ac.;* of outer ear, Petrol.; sensation of something FALLING before ear, Cocc. (r.); FLUTTERING, Mag. c., Spig.; HAMMERING, Croc., Pso., Sil.; HEAT and burning in and about the ear, Jac.; of external ears, Mur. ac.; heaviness in and before the ears, CARB. V.; habitual HERPETIC ERUPTION in meatus, Grap.; HISSING, Dig.; HUMMING, Nicc., Sec. c.; ITCHING in ear, Bov.; and suppuration, *Am. car.*, Bov.; LANCINATION, Con.; sensation of something LYING before ear, Mag. m., Sul. ac.; NOISES, *Chin.*, Coca, Merc., *Pul.;* and digging pain extending to nostril, Jac.; musical noise, fear of apoplexy, great noise in the ear, distraction and loss of memory, Cannab.; NUMB sensation about ear (after Sul.), Lach.; OTITIS, Merc., Thu.; PAIN, *Asar.*, Ig., and chilliness, *Bell.*, *Cham.*, *Merc.*, PUL.; pain and ill humor, Nx. v.; aching PRESSING, Ip.; RINGING, Arg. n., *Grap.*, Led., Pho., Pso.; fluttering, Bell., Mag. c., Rhe.; ROARING, Aur., Bor., Croc., Pho., Sil., Thu.; RUSHING, Calc. ostr.; noise like a SEA-SHELL (after Sul.), Lach.; SENSITIVENESS to noise, Merc.; sound, Am. car., Arn., Lyc., Merc., Plat., Sul.; SINGING, Pso.; SNUFFING (in children), Kre.; loud SOUNDING, Merc.; SPATTERING noise, Plat.; STINGING, Am. mur.; STOPPAGE, Ars., Asar. (l.), Calc. ostr., Coc. c., Con., Gel., Grap., Hydrs., Iod., Lach., Nit. ac., Phyt., *Pul.*, Sil., Spig., Sul.; SWELLING, > removing wax, Calc. ostr.; THROBBING, Hep.; TINGLING, Sul.; large ULCERS, Led.; black WAX, Pul.; dry (after Sul.), Lach., Nit. ac.; hard, Sel.; increased, Mur. ac., Zn. o.; like chewed paper, Con.; WHIZZING, Mag. c.; sound of WIND, Plat.; imperfect ACTIVITY, alternating with absence of mind, Alum.; anxiety and sensitiveness to noise, Sil., Pul.; hypochondriac humor, confusion of mind, Agn., Bap., Carb. a.; insanity, sensitiveness only after loud screaming, half conscious, Ars.; dull dreamy state, Zn.; want of memory, Mos.; mistakes in speaking, Bov.; HEADACHE, Bar. m., Chin., Sul.; stitches in head, Crot. t. (r.); heavy pressure and heat on vertex, extending to both ears, with soreness of brain, Sul.; old look to FACE, Kre.; numb sensation down to the cheek, Lach.; wedge-shaped TEETH, Kre.; swelling and induration of the TONSILS, Merc., Nit. ac., Stap., after abuse of mercury, Nit. ac., Stap.; burning in STOMACH, Bar. m.; hunger daily at 11 A.M., Lach.; attacks of vomiting, Kre.; CONSTIPATION, Sul.; TETTERS, Dul., Fer., Nicc.; TREMBLING, Pul.; FAINT feeling at 10, 11 A.M., *Sul.;* typhoid FEVER, Apis, Ars., Bell., Bry., Carb. v., Hyos., Lachn., Lach., Merc., Nit. d. s., Pho., Pho. ac, Pso., Pul., Sec. c., Stram., Sul.; hot stage, Lachn., *Rhu. t.;* hot flushes in face followed by cold sweat, Sul.; burning of soles at night, Sul.; cold stage, *Cham.*, Cinch., Pul., Rhu. t.; sweat on back, Pul.

IMPAIRED. ALTERNATING with acute, Anac.; with noises, Anac.; eye symptoms, Guare.

—CLOSED. See PLUGGED.

—DISTANT sounds, for, Pho.

IMPAIRED. FALLING before the ear, as if, Calc. ostr., Nit. ac., Sul.; falling into the ear, Con.
— HAND were held over ear, as if, Chel.
— IMPRESSION, hears every thing but it makes no, Hell.
— INTERMITTENT, Sil.
— INTOXICATION, as in slight, Op.; does not answer questions, Mag. c.
— LEAF, lying before, as if, Ant. cr., Sul. ac.
— LYING before, as if something were, Acon., Agn., ASAR., Bell., Carb. cr., Cannab. s., Carb. v., Cinch., COCC., Coloc., Cyc., Hyos., LED., Mag. m., Meny., Par., Pho., Rhe., Rhu. t., Saba., SPIG., Sul. ac., Verb.
— MIST, like a, Par.
— NOISE, as through a great, Ol. an.
— OPENED wide and hollow, as if ear were, Con.
— PART, hears a, Mang., Spig.
— PERIODICAL, Spig.; every other day, Pho., Pul.; every week, Sul.
— PLUGGED (includes CLOSED, OBSTRUCTED, etc.), as if, Acon., *Æth.*, Agn., Am. car., Anac., Angu., Ant. cr., Arg., *Arg. n.*, Ars., *Asar.*, Bell., Bism., Bor., BRY., Cala., *Calc. ostr.*, Cannab. s, *Carb. v.*, CAUS., Cham., CHEL., Cinch., Cocc., Colch., *Coloc.*, CON., *Cyc.*, Gel., *Grap.*, Guai., Hyos., *Jac.*, IOD., *K. ca.*, *Lach.*, *Lachn.*, *Led.*, LYC., *Mag. m.*, *Manc.*, MANG., Merc., MENY., MERC., *Mez.*, *Na. c.*, NIT. AC., NX. m., *Par.*, *Petrol.*, Pho., *Phyt.*, PUL, Rhe., Rhu. t., Rum. c., Saba., SEL., SEP., SIL., SPIG., Spo., Stan., SUL., Sul. ac., Syph., Tel., VERAT., VERB.; WITH ringing, Mez.; roaring, Grap., Merc., Seneg., Sep.; sensation as if ear-wax were running into mouth, Lach.
— SKIN were stretched over ear, as if, *Asar.*, with ringing and roaring, Cannab. s.
— SUDDEN, Con., Dig., Gel., Nicc., Pb.
— VOICE, for human, Mur. ac., Pho., *Sil.*, Sul., Sul. (l.); especially for the human voice, Mur. ac., *Pho.*, Rhu. t., Sil., Sul.; for his own voice, Coca, Pho.; for every thing except human voice, Ars., Ign.; with sensation of wind in ear, Ign.
— WATER RUSHING, as from, Cocc.
— WATCH, for, *Pho. ac.*

IRRITATION of. See ACUTE.

LOST (including DEAFNESS), *Acon.*, Æth., Aga. m. (l.), Aloe, Alum., *Ambra*, *Am. car.*, Am. cau., *Am. m.*, Amyg., *Anac.*, Angu., ANT. CR., Ant. t., *Apoc.*, *Arg.*, *Arg. n.*, *Arn.*, ARS., Asaf., ASAR., *Aur.*, *Aur. m.*, *Bap.*, *Bar. c.*, *Bar. m.*, BELL., Bon., *Bor.*, Bov., *Bry.*, *Bursa par.*, *Cact.*, CALC. OSTR., Cannab. i., *Cap.*, Carbn. s., Carbn. s. (l.), Carb. v., Carl., CAUS., *Cham.*, CHEL., Chin. s., Chin. s. (l.), *Cinch.*, *Cist.*, *Coca*, *Com.*, CON., Cot. (l.), Croc. c., Crot. h., *Cup.*, Cup. ac., Cyc., *Dig.*, *Dro.*, DUL., *Elap.*, Eth., Eup. pur., Gas., GEL., *Glo.*, GRAP., *Hydrs.*, HYOS., *Ig.*, Ip., *Iod.*, Jat., K. ar., *K. ca.*, K. n., KRE., *Lach.*, Lau., LED., Lepi., Lepi. (l.), Lol., Lon., LYC., *Mag. c.*, *Mag. m.*, MANG., MENY., MERC., Merc. c., Merc. m., MUR. AC., *Na. c.*, NA. M., Na. sal., *Nicc.*, *Nitr.*, NIT. AC., NX. M., Ol. an., Olnd., OP., PETROL., PHO., PHO. AC., Phyt., *Pb.*, PUL., Pul. n., Rap., *Rhodo.*, *Rhu. t.*, *Rhu. v.*, Rut., Saba., Saln., Sal. ac., SEC. C., *Sel.*, *Sep.*, SIL., SPIG., Stan.,

Stap., *Stram.*, SUL., Sul. ac., *Tel.*, *Verat.*, (Verat. v.), *Verb.*, Vip, Zn.; MORNING, after rising, Stan. (l.); 10 A.M., during vibration in head, Grat.; AFTERNOON, Sil.; 8 P.M., Nicc.; after CONVULSIONS, Sec. c.; after rough COUGH, Led.; during DINNER, Sul., and afterwards, Carbn. s. (l.); after HICCOUGH, Bell.; during MENSES, Lyc.; on blowing NOSE, Spig.; after ROARING, Sep.; on WAKING, Oena.; (AGG.) EVENING, Bell.; 9 P.M., on lying down, Merc. c.; (AMEL.) boring in FINGER, Spig.; blowing NOSE, Stan. (l.); RIDING in a wagon, Grap.; during SENSITIVENESS to sounds of vehicles, Chen. a.; WITH, howling, Con., Kre., Sil.; itching and fetid discharge, Bov.; itching and suppuration, Am. car.; sensation of a leaflet before ear, Ant. cr. (r.); pain, Bry., Cyc.; ringing, Arg. n., Chin. s., Con., Sul.; shooting, Bell.; sensation of something placed before ears, Bell., Led.; stopped sensation, Cala., Mang., Peti. (r.), Sep., Spig. (l.); scabby eczema behind ear, Pso. (r.); sensation of wind in ears, Cocc. (r.).

DEAF AND DUMB. Calc. ostr., Sep., Sil., Sul.; stupefaction, Am., Carb. a., Crot. h., Kre., Ol. an., Stram.; dull headache, great anxiety, trembling and sweating, Chin. s.; buzzing in head, Pso.; vertigo, Crot. h.; catarrh, Lach.; congestion of blood, Bell.; eructations, Petrol.; anxiety in abdomen, Aloe; pain in abdomen, Crot. t.; cough, Chel.; as if some one HELD HAND over the ear, when coughing, Chel. (r.); as if ear were wide OPEN internally and HOLLOW, after ringing in ear, Aur. m.; SUDDEN, Scr.; with stopped sensation, when walking, Cic. v.; temporary, as if something had fallen into the ear, with lancinating pains, principally coming on after blowing the nose, Chin., Con.; THREATENING, Olnd., Pul.; SENSATION of, Bar. c., Coca; <lying on ear, Coc. c. (l.).

LOW TONES of music incorrect, Bry.

REVERBERATING (including ECHOING, RE-ECHOING, and RESOUNDING), Alum., Bar. c., Carb. an., *Caus.*, Eth., GRAP., Iod., Mag. m, Merc., Merc. sol., NIT. AC., NX. V., PHO., PHO. AC., Pul., Rhodo., Spig, Ther.; MORNING, Caus., Chin. s. (r.), Lyc. (r.), *Pho.;* on BLOWING nose, *Bar. c.:* <MORNING, Caus.; EATING, or after it, Nx. v.; >WALKING, Cop.; LONG, of strong tones, Rhodo.; of MUSIC, Nx. v., Pho. ac., Pul., Spig.; of STEPS, Caus.; STRONG, Nx. v.; of WORDS, CAUS., Gas., *Pho.;* in morning, >after dinner, Nx. v.; on waking, Pul.; of every word and every step, with hard hearing, Caus.; of his own voice, Nit. ac.; his own and others, in morning, Pho.; of his own, and it sounds like dumb-bells, Spig.; ASCENDING by octaves, >after breakfast, Ant. cr.

SENSITIVE. See ACUTE.

VIBRATING of every sound, Pho. ac.; of loud speaking, Pho.; violent, or the sounds go through the ear, with hard hearing, Merc.

VOICES sound like a HUMMING, Bell.; his own voice sounds like DUMB-BELLS, and resounds through the head, Spig.; STRANGE, his own seems, Alum., Alum. (r.), Rum. c.; with sounding and humming in left ear as of a sea-shell, cannot tell where the person is who speaks unless he sees him, from the right to left ear talking loud is very painful, Terb. m. m.; WHISPERING, Nit. ox.

WATCH sounds like a hissing, Pho. ac.

INTERNAL EAR. SUBJECTIVE SOUNDS, Am. car. (l.), Ars., Benz. n., Meny.

(r.), *Sep.*, Spig.; in FORENOON, >boring with finger, Castor.; WITH vertigo, Sang.; PENETRATING whole body, with a sensation of a wind in ear, Ther.

AIR, escaped from ear, as if cold, when laughing, Mil.

ARTERIES beating, cerebral, Op.

BAGPIPE, distant, when lying on side, >rising, Na. c.

BANGING, Bar. c., Na. c., Nit. ac., Saba., Zn.

BAT, Mil. (l.); at night, Pho. ac.

BEATING, Alum., Am. car., Am. m., Bar. c., Berb., Bro., Calc. ostr., Carb. v., Caus., Cinch., Coloc., Con., Ign., K. ca., Lyc., Na. c., *Na. m.*, Nicc., Petrol., Pho., Rhodo., Sang., Sul.; WITH buzzing, Nit. ac.; clapping, Na. m.; humming, Mur. ac., Nit. ac.; DISTANT, >rising, Mez. (l.); as if against a DOOR, Ant. cr.; PULSE-LIKE, Bar. c., Caus., Coca, Coloc., Dig., Glo., Grap., Lach., Mag. m., Merc., Merc. c., Sep., Sil., Sul. ac., Zn.

BED, as if some one were under the, Bell., with a rattling noise, Canth.; over, Calc. ostr.

BELLS (compare with RINGING, TINKLING, and TOLLING), Am. car., Arun., Gas., Hyos., *Led.*, Na. slfc. (r.), Pho. ac.; 10 P.M., Val.; when WALKING, Chin. s. (r.); while YAWNING, Mez. (r.); < NIGHT, lying down, Sil.; CHURCH, with mania, anguish, and anxious sweat, Ars.; CLEAR, Sul. ac. (r.); DISTANT, Coff. t., Der. p., Na. slfc. (l.); MIDDLE-TONED, Coca.

BLOOD streaming through brain, Con., Op.; rushing to brain, with a dull noise, at every beat of heart, Op.

BLOWING, Hydrc., Ox. ac.; whizzing, Chel.; then ringing, Na. m.

BOUNCING, rattling, confused falling of hard things, Sep.

BUBBLING, Con., Dul., Lyc., Mag. c.; WITH coldness and dryness of ear, Berb.; dulness of sensorium, Plat.; BEFORE ear, morning after waking, Bell.; as of something FALLING before and then away from, on becoming erect, and leaning back, Grap.; as of a liquid, Thu. (r.).

BURSTING, like a bubble, Grap., Lyc., Sul.; then buzzing, Carb. b.; as of something fallen upon floor, Saba.; as if something were falling to and fro, Saba.

BUZZING, Acon., Alco., Aloe, Alum., Am. car., *Arg. n.*, Arn., Bell., Cact., Cai., *Calc. ostr*, Canch., Cannab. i., *Canth.*, Carbn. s., Carl., *Caus.*, Chel., Chen., a., Chin. s., Coff., *Con.*, Cop., *Dro.*, *Elap.*, Elap. (r.), Gam., Glo., *Grap.*, Hcy. ac., Hydrphb., Hyos., Iod., Ir. v., K. iod., Kalm., Lyc., Mag. c. (r.), Mag. m., Manc., Merc., Murx., Nicc., *Nit. ac.*, Nx. m., Nx. v., Pic. ac., Pho, Pso. (l.), Rhodo., Ric., Saba., Spig., Sul., Sul. ac., Sul. iod., *Tarent.*, Thu. (r.); in MORNING, Dios.; FORENOON, <whistling, Rhodo. (l.); NOON, Ced., Fago.; AFTERNOON and EVENING, <after sensation of a leaflet bursting, Gam.; evening, *Bar. c.*; 2.30 P.M., > 4 P.M., Murx.; 7 P.M., Physo.; 10 P.M., Ham.; when DESCENDING STAIRS, Crot. c.; on SWALLOWING, Rhodo.; < SITTING, > LYING, STANDING, and WALKING, Bell.; < BURSTING of a bubble, Carb. b.; WITH hard hearing, Bov., Mag. c., Petrol., and whizzing, Lyc, Merc.; growling, after roaring, Bell.; humming and illusions, Mag. m.; itching and roaring, Sep.; inflammation, Merc.; dry earwax, Pho.; difficult comprehension, Ars.; stupor, Pso., and dulness, Lach.; FOLLOWED by clucking, Aga. m.; like BEES, FLIES, Am. car., Con., Elap., Mag. c., Nit. m., Nx. v., Sal. ac.; BEFORE ears, Am. car., Bar. c., Mag. m., Sol. n.; FINE, with headache, Glo. (l.); HUMMING, Carb. ac.; with hard hearing, Cact., *Dro.*, Iod., Nicc., Sec. c.; RUSHING, on stooping, Mang.

CALL, sudden waking from a, Ars.

CANNONADING, Bad., Chel., Mos. (r.); then a few drops of blood come out, Mos.

CASCADE, Rhu. t.

CAT SPITTING, in afternoon, Nit. ac.

CHIRPING, Rat., Rat. (r.); at night, Mur. ac.; like CRICKETS, Carb. v., Caus., Euphm., Euphm. (r.), Fer., Meny., Nicc., Nicc. (l.), Sil.; before ears, Fer.; like a GRASSHOPPER, Carb. v., Nx. v., Tarax., Tarax. (l.); like LOCUSTS, at night, Nx. v.

CLANGING, SIL., *Spig.;* like a musical, monotonous, melancholy song of the water-toad in Germany, Mang.

CLAPPING, Grap., K. ca., Na. c., Rhu. t., Saba., Sil., Stap., Zn.; with cracks, fright and starting, Rhu. t.; knocking, Pho.

CLICKING, spasmodic, opening and shutting (like closing or opening the fist), very annoying, Nitr.

CLINKING, while scratching occiput, Am. n. (l.).

CLIPPING, Grap. (r.).

CLOCK, Ter.; in morning, Mang.

CLUCKING, Aga. m. (r.), Bar. c., Cad. s., Grap., Lyc., Petrol., Rhe., Sil.; after BUZZING, Aga. m.; on STOOPING, Grap.; RISING from stooping, Sep.; WITH noise as if something were falling to and fro, Grap.; heaviness of head, Grap.

CONFUSED, Carbn. o., Fago., Hydrc. (l.), Par.

CRACKING, Bar. c., Calc. c. (l.), Cocc., Ery. a. (l.), Grap., K. ca., Lach. (r.), Mos., Na. m., Na. m. (l.), Nit. ac., Nx. v., Ol. an., Petrol., Saba, Sang. (r.), Stry., Sul., Tarent. (r.), Thu.; in MORNING, Na. c. (r.); in bed, on moving jaw, *Grap.;* EVENING, Petrol.; when eating, Grap.; 2 P.M., when eating, Ped.; after BREAKFAST, Zn.; when CHEWING, Calc. c., Calc. ostr., Mang., Meny., Na. m., *Nit. ac.*, Sil.; BLOWING NOSE, Pry., Hep.; when COUGHING, Nx. v.; moving HEAD, Grap. (r.), *Pul.;* READING aloud, Aloe; on SNEEZING, Bar. c., Bry., Meny.; STROKING cheek with finger-tips, (Sang.) (r.); on SWALLOWING, Bar. c., Calc. ostr., Cic. v. (r.), Coca, Coc. c. (l.), Der., *Elap.*, Grap. (l.), Mang., Na. m. (l.), Sil.; on WALKING fast, Bar. c.; WITH opening of the ear, Mar., Mur. ac.; claps, fright and starting, Rhu. t.; pinching, Na. c.; swashing, on swallowing, Grap.; BEFORE ear, Lach.; INTERMITTENT, *Petrol.;* LOW, FLAT, SLIDING, Grap.

CRACKLING (includes CREPITATION), Alum., (Ambra), Aur., Aur. (l.), Bar. c., Bor. (l.), *Calc. c.*, Calc. ostr., Coc. c. (r.), Dul., Dul. (l.), Eup. per., Eup. pur., Glo., Glo. (l.), Grap., *K. ca.*, Lach., Lachn., Mos. (r.), Nit. ac., Rhe, Saba., Sep., Spig.; EVENING while sitting, Hip. (l.); BLOWING nose, Bar. c. Calc. ostr., Mar. (r.); EATING, Calc. ostr., Grap., K. ca., Mang., Meny., Na. m., Nit. ac., Ped. (r.), Sil.; when LYING upon ear, Bar. c. (l., then r.); on SWALLOWING, Alum., *Bar. c.*, Calc. ostr., Elap., Eup. pur., Mang.; on WALKING, Nit. ac.; WITH an indescribable unpleasant sensation, Mos.; pain, on eating, Na. m.; BEFORE ears, as from rattling of paper, Sep.; as from ELECTRIC SPARKS, Hep.; EXTENDING into forehead, > holding hand over eye, Spig.; as from STRAW, on every motion of jaw, Carb. v.

CRASHING, Bar. c., Cic. v., Cocc., Grap., Hep., Rhu. t., Saba.; 10 P.M., Con. (r.); night, Bar. c.; in BED, Grap.; on moving jaw, CHEWING, eating, Calc. ostr., Grap., K. ca., Mang., Meny., Na. m., Sil.; when BLOWING NOSE, Mang.; when READING aloud, Aloe; when SNEEZING, Bar. c., Meny.; when SWALLOWING, Bar. c., Coc. c., Meny.; when WALKING fast, Bar. c.; WITH pain, Na. m.; amelioration of hard hearing, Grap.; as from breaking a pane of GLASS, on falling asleep, Zn; like a distant SHOT, Am. car., Chcl., Plat.

CREAKING, Grap., Pul., Stan., Thu.; EVENING, Stan. (l.); on BLOWING NOSE, hawking, etc., Bry.; when SWALLOWING, Aga. m., Grap. (l.), Thu.; when lying on affected side, WITH pulsation in ear, and pimples and pustules in external ear, Spo.; BEFORE ear, in evening, Stan. (l.); as from FROGS, while sitting, Mag. s. (l.); when walking, Mang.

CREPITATION. See CRACKLING.

CRYING aloud in a dream, hears himself, Bell.

CYMBALS and DRUMS, Lol.

DETONATION, Cic. v. (r.), Itu., Mos. (r.), Grap. (l.); on blowing nose, Hep.; on swallowing, Bar. c., Bar. m., Cic. v.; like shocks, as of a cannon, Bad., Mos.

DRUM would burst, Rhu. t.; had burst, during siesta, Rhu. t. (l.); as if something struck, Sil.

DRUMMING, Bell., Canth., Cup., Dul., Dul. (l.), LACH.; on WALKING, Manc.; WITH SOUNDS of cymbals, Lol., and trumpets, followed by roaring, Bell.; DISTANT, Dro.; when lying on ear, > rising, *Cup.*; DULL, Bor. (l.); KETTLE-DRUMS, Bell.; as on the top of a vaulted ROOF, Bor.

DULL (dead), opening with, after a closed sensation, Mar.

EXPLOSION as from the breaking of glass, after getting into bed, Aloe.

FLAPPING, regular, Sil.; like the wings of a large bird, Mos., Olnd., Plat., Spig.; with a discharge at every step as if a valve were opened and shut, Grap.

FLUTTERING, Bell., Calc. ostr., Carl., Cup., Cup. (l.), Mag. c. (r.), Mag. m., Mag. m. (r.), Merc., Merc. d. (l.), Merc. sol. (l.), Mur. ac., Nit. ac., Plat., Pul. n. (r.), Sel., Sil., Spig., Sul.; at 11 A.M., Mag. c. (r.); WITH hard hearing, Mag. c., Spig.; warmth, Mang.; numbness as if drunk, in open air, does not understand what he is asked, Mag. c.; BEFORE ears, Mang. (l.), Merc. sol. (l.); morning after waking, Bell.; 5 P.M., Sul. (l.); as from a bird, Ant. t. (l.); evening, Mag. c. (r.), Tab.; RHYTHMICAL, Sil. (l.); like WINGS, Cham., Jac., Mag. c., Mag. c. (r.), Pho. ac.; a bat's wing, Mil.; a butterfly's, Na. m.; during dinner, Na. m. (l.).

GROANING, Thu. (r.).

GROWLING (like a bear), Anac., Am. m., Aur., Bell., Bry., Caus., Con., Kre., Lach., Lyc., Na. c., Na. m., Nit. ac., Nx. v., Pul., Saba., Sep., Sil., Spig., Sul.; with buzzing, after roaring, Bell.; followed by ringing and hissing, Kre.

GRUMBLING, Sil.; night, Am. m. (r.).

GURGLING, Aga. m., Ammc., Bar. c., Bell., Berb., Caus., Dul., Grap., K. ca., Lact., Lau., Lyc, Mag. c., Mur. ac., Olnd., Pho., Plat., Sang., Sep., Sil., Zn.; WITH loud noises, Rhodo.; SYNCHRONOUS with the pulse, Merc. c.; as if

WATER were in the ears, Rhodo., Spig., Sul.; as if water were running from a bottle or down the gullet, Grap.

GRUNTING, on swallowing, Calc. c.

GUNS, reports of, Cannab. i.

HAMMERING, Spig., Thu.; with hard hearing, Croc., Pso., Sil.; coldness of whole body, Thu.; much urination, Thu.

HISSING, Acon., Æth., Alum., Cai., Calc. c. (r.), Chel., Chin. s., Coc. c., Dig., Gam., Grap., Hep., Ill., K. ca., Kre., Led., Lyc., Mag. c., *Mar.*, Mur. ac., Na. s. (l.), *Nx. v.*, Pic. ac., *Sil.*, Sum.; MORNING, from forcibly drawing air in nose, from eructations and passing hand through hair, Mar. (r.); EVENING, Calc. c.; after forcibly INHALING air through nose and when TALKING, Mar.; WITH hard hearing, Dig.; ringing, after growling, Kre.; sensation of wind in ears, Dig.; RINGING, *Nx. v.;* as from boiling WATER, Bry. (l.), *Dig.*

HOWLING, with deafness, Con., Kre., Sil.; deep-toned, and synchronous with pulse, Sep.

HUMMING, Acon., All. s., Am. m., Amyg., *Anac.* (l.), Ant. cr., Arn., Aur., *Bell.*, Benz. ac., Benz. ac. (l.), Bry., Bry. (l.), Calc. c., Calc. ostr., Canth., Carb. ac., Carb. v., Carbn. s. (l.), Card. b., Carl., Chel., Cinch., Cob. (l.), Coca, Cop., Con., Croc., Der., Dro., Fer., Feru., Fl. ac., Gam., Gel., Glo., *Grap.*, Jal., K. ca., K. cl., Kre., Lach., Lyc., *Lyc.* (r.), Mag. m., Merc. (r.), Mez. (r.), Mur. ac., Na. m., Nicc., Nit. ac., Ol. an., Op., Pho., Pb., *Pul.*, Ric., Saba., Sang., Sec. c., Seneg., *Sep.*, Sil., Spig., Stry., Sul., Tab., *Verat. v.*, Zing.; MORNING after rising, Ars.; on waking, Na. m., Rhodo.; 11 A.M., Zing.; EVENING, Sep.; on forcing AIR into it, before dinner, Saba.; after ITCHING, Na. m.; LYING down, All. c.; TALKING or WHISTLING, Op.; (AGG.) loud noise on going into open air, Tab.; when sitting, Bell.; (AMEL.) laying head on table, Fer.; lying, standing, or walking, Bell.; WITH dull hearing, Nicc., Sec. c.; roaring, Con. (r. and l.), on stooping, Croc.; roaring, also in the head, Caus.; sensation as if something heavy fell and burst, then loud ringing, Saba.; heaviness in head, *Ars.;* lying in a stupor, Carbn. o.; FOLLOWED by itching, Na. m.; like BEES, Art. ab., Nx. v., Sal. ac.; BEFORE ears, Aur. (l.), Carb. v., Dro., Kalm., Kre. (l.), Lact, Mag. m., Rhodo.; DULL, morning on rising, Sil.; EXTENDING to back of head, Carbn. s. (l.), Spig. (l.); as if about to FAINT, Merc.; as from INSECTS, Meny.; like a SPINNING-WHEEL, Aga. m.; as if something were STICKING in ear, Merc. sol. (l.); like WASPS, Merc.

KNOCKING, Nit. ac.; and clapping, Pho.; out of doors and some one calling him, in a dream, Ant. cr.

LOUD, Mag. s. (l.), Sul.; after dinner, Mag. c. (r.); in a dream, Stan.; and a pistol-shot, in a dream, waking him, Cerv.; with hard hearing, Merc.

MACHINERY, 2 P.M., Hydrs.

MILL, Cic. v. (l.), Iod., Nx. v., Petrol.; at a distance, Mez. (l.), Bry. (l., then r.); going in the head, morning on waking, Naj.

MURMURING, Bell., Hep. (l.), Sil.; AFTER SLEEP, $<$ mental anxiety, Act.; BEFORE ear, in evening after lying down, Hep.; RHYTHMICAL with pulse, Carl., Pul., Sec. c.

MUSICAL, Na. c., Pul., Sal. ac., Sep., Sil.; evening on lying down, Pul.; after a dream that he heard music, Sarr.; in the evening, the music that she heard in the daytime, Lyc.; with diminished hearing, fear of apoplexy, great noise in the ear, distraction, and loss of memory, Cannab.

NAIL driven into a board at a distance, Aga. m.

NOISES, Acon., Arun., *Bell.*, Bol. s., Calc. c. (l.), Calc. ph., Camph., Carbn. o., Carbn. s. (l.), Chel. (l.), Croc., Coff. t., Dig., Euphm. (r.), Hydrs., Hyos., K. ca., Lyc., Mag. s. (l.), Na. m., Na. s., Nit. ox., Pho., Rhu. v. (r.), Sep. (r.), Spig., Stry. (l.), Sul., Tab., Tarent., Tep., Verat. v., Zn.; MORNING, Alum., Arg. n., Aur., Bell., Calc. ostr., *Fl. ac.*, Gam., Grap., Lach., Lyc., Mag. c., Mang, Merc., Mez., Na. c., Na. m., Phel., Plat., Pul., Rhodo, Sil., Sul., Tab., Val., Zn.; in bed, Mag. c. (l.); 8 A.M., Phe.; AFTERNOON, Ambra, Ant. cr., Carb. v., Nit. ac., Pul., Rhu. t.; EVENING, Alum., Bar. c., Canth., Caus., Grap., Lact., Lyc., Mag. c., Merc., Nicc., Ol. an., Petrol., Pho. ac., Plat., Rhodo., Sep., Sul., Sul. ac., Zn.; NIGHT, Am. car., Am. m., Bar. c., Carb. a., Con., Croc., Euphm., Grap., Hep., Mur. ac., Nx. v., Pho. ac., Rat., Sep., Sil., Thu., Zn.; on waking, Hydrs.; MIDNIGHT on waking, Rat.; after repeated small doses of ALCOHOL, Ars.; in BED, Tarent (< r.); on BLOWING NOSE, Carb. a., Lyc., Mar., Meny.; BORING in with finger, Castor., Chel, Lach., Nicc.; on CHEWING, Alum., Ant. cr., Carb. v.; on EATING, Con., Grap., Na. m., Sul., Sul. ac.; after eating, Canth., Cinnb., Con., Grap., Mag. c., Op., Sil., Zn.; on waking after DREAMS, Bell., Grap., Lach., Na. m., Nx. v., Rat., Sep.; in DOORS, Cic. v., Mag. c.; out of doors, Aga. m., Benz. ac., Carb. a., Mag. c., Tab.; after mental EXERTION, Con.; during FULL MOON, Grap.; on OPENING MOUTH, Dul., Sul. ac.; PASSING finger over cheek, Sang.; hand over ear, Mar.; on RUBBING or SCRATCHING, Meny.; on going to SLEEP, Zn.; on SWALLOWING, Alum., Benz. ac., Cic. v., Coc. c., Grap., Lepi., *Rhe.*, *Sil.*; on empty swallowing, Thu.; on quick WALKING, Pho.; after WRITING, Sep., Zn.; (AGG.) after eating, Carbn. s. (l.); on waking, Tarent. (< r.); (AMEL.) after headache, Tarent.; WITH, or alternating with ear symptoms, Anac., Arn., Bell., Carb. v., Chin., Lach., Lyc., Merc., Nit. ac., Nx. v., Pho., Pul., Sep., Spig., Sul., Zn.; WITH digging pain extending to nostril, and hard hearing, Jac.; discharge, Calc. ostr.; deafness, dull hearing, Merc., so that music nauseates, Sul.; rushing, and roaring in head, Lach.; anxiety, Pul., Sil.; insanity, Ars.; desire to be alone, Con., and at rest, Bell.; hatred of company, Bell., Pho.; distraction of mind, and loss of memory, Camph.; hypochondriac mood, anxiety, Pul.; hypochondriac humor, and confusion of mind, Agn., Bap., Carb. a.; want of memory, Mos.; starting, frightened, Mil., when falling asleep, Sul.; vertigo, Tep.; dimness of vision, K. bi.; red eyes, Glo.; tears and coryza, Bry.; belching, Carb. a., *Grap.*; sleeplessness, Bar. c.; horripilation, Alum.; ALTERNATING with dull hearing, Anac.; SUDDEN, Ast., Mur. ac.

PIPING, Bor. (r.), Lyc.

QUIVERING, Bov., K. iod., Spig.

PULSATION of carotids, Verat. (l.).

RAILROAD TRAIN in a tunnel, Nit. ox.

RAIN, (Bov.); striking the ground, Rhu. t. (l.).

RATTLING, with noise as if some one were under the bed, Canth.

REPORTS, Eup. pur., Nit. ac., Stap.

RINGING (compare with BELLS), *Acon. c.*, *Acon.*, Aga. m. (l.), *Agn.*,

Alco., Ail. (r.), All. c., Aloe, *Ambra*, *Am. car.*, *Am. m.*, Anac., *Angu.*, Angu. (r.), Ant. cr., *Arg. n.*, *Arn.*, Arn. (l.), *Ars.*, ASAF., *Asar.*, Ast., Atro., AUR., Aur. m, *Bar. c.*, BELL., Bor. (r.), Bro. (r.), *Bry.*, Brucn., Calc. cau., Calc. c., CALC. OSTR., Calth., *Camph.*, CANNAB. I., Cannab. s., *Canth.*, *Cap.*, *Carb. a.*, Carbn. o., Carbn. s., CARB. V., *Carl.*, Castor., CAUS., Caus. (l.), CHAM., Cham. (r.), *Chel.*, CHIN., CHIN. S. (< l.), Chlf., Chlo. (r.), Chlol., *Cic. v.*, Cic. v. (l.), *Cinch.*, *Cit. v.*, *Cle.*, Coc. c. (l.), Coca, Cod., Coff., Colch., Colch. (r.), Coloc., Com., *Con.*, Con. (r.), *Croc.*, Cup. ac., Cyc., Dig., *Dul.*, Elap., Ery. a. (l.), EUPHB., *Euphm.*, *Fer.*, Fer. (r.), Gam., Gas., Gin., GLO., *Gran.*, *Grap.*, Grap. (l.), Guare., Hell., Hell. v., HEP., *Hyos.*, Hydrc., Hydrs., Hur., *Ig.*, Ill., Iod., *K. ca.*, K. cy., K. bi., *K. iod.*, K. n., Kis., *Kre.*, LACH., Lachn., *Led.*, LYC., Lyc. (r.), MAG. C., *Mag. s.*, Mag. s. (l.), Manc., *Mang.*, *Mar.*, Meny. (r.), MERC., Merc. cy., Merc. n., *Merc. sol.*, Mez., Mil. (r.), Mor., MUR. AC., Myric. (l.), Na. c., Na. m., Na. sa., Na. slfc. (r.), Nicc., *Nitr.*, *Nx. v.*, OLND., Ol. an., *Op*, Osm., Paeo., *Par.*, Pb., Pb. cr., Pen., *Petrol.*, Phel., *Pho.* (< r.), *Pho. ac.*, Plan., PLAT., Pso., Pte., PUL., *Pyrus*, *Rat.*, Rhodo., RHU. T., Rhu. v. (r.), Rum. c., Rut., Saba., Sal. ac., Saln., SARS., Sars. (l.), Scr., Sec. c., Sel., *Sil.*, SPIG., Sol. t. æ., *Spo.*, Spo. (r.), *Stan.*, Stan. (l.), Stram. (< l.), *Stap.*, Stap. (l.), *Sul.*, Sul. (r. and l.), Sul. ac., Sulphs. ac., Tab., Tana., Tarax., Tarent., Ter., Thu. (r.), Til., Val., *Verat.*, *Verat. v.*, Vinc., Vinc. min., *Vio. o.*, Xan. (< r.), *Zn.*, Zn. (r.); DAY, Sul.; MORNING, Cle.; in bed, Arg. n., Sul.; on rising, Mez. (r.); FORENOON, Carb. v., Fl. ac. (r.); 2 A.M., Chin. s.; 9 A.M., before headache, > cold water, Euphr. (l.); during coldness, Chin. s.; 11 A.M., Na. m.; NOON, Glo. (l.); AFTERNOON, Carb. v. (l.), Kalm., Sul. (r.); when writing, Carl. (r.); EVENING, K. n., Sil.; in bed, Croc., Pho., Val.; 2 P.M., Verat. v.; 3 P.M., Fago. (l.); 4 P.M., Dios. (< r.); 5.45 P.M., Ol. an.; 7 P.M., Physo.; 8 P.M., Ham.; NIGHT, Carb. a., Cyc., Pho. ac., Zn., Zn. (r.); MIDNIGHT, on waking, Rat.; on going into OPEN AIR, Carb. a. (r.); BORING in with finger, Chel.; during DINNER, *Sul.*; after dinner, Mag. c. (r.); in a DREAM, it wakens him, and he hardly believes it to be a dream, Zn. c.; while EATING, Sul.; on closing EYES, Chel.; moving HEAD, Stap.; BLOWING NOSE, Mar. (r.); after a NOISE like a blow inside, Na. m.; after SHOCKS and contractions in ears, Na. m.; when SITTING, Ars. (r.), Merc. cy., Sul.; on SNEEZING, Euphm.; after STOOL, Apoc. c. (r.); when TALKING, Spig.; on WAKING, Tarent. (r.); when WALKING, Chel. (l.), Manc., Nicc., Rhu. t.; in open air, Aga. m. (r.); (AGG.) evening, *Merc. sol.*; night, with headache, Cyc.; after lying down, *Sul.*; (AMEL.) cold water, Euphr.; digging into, Nicc.; during rest, Stap.; on rising, Tarent. (r.); rubbing, Meny. (r.); WITH deafness, Con., Sul.; hard hearing, Arg. n., *Grap.*, Led., Pso., as if plugged up, Mez.; hissing, after the growling, Kre.; roaring and dull hearing as if a skin were stretched over the ear, Cannab. s.; roaring, tearing, and twitching, from one's own singing, Pul.; whistling, Vin. min.; crawling tickling, Cinch.; illusions, Val.; headache, Acon. c., Carbn. s., Cinch., Euphr., Gin.; vertigo, Carb. v., cold stage, CHIN., GRAP., Rhu. t.; FOLLOWED by hard hearing, Aur. m.; ALTERNATING with tickling itching in right Eustachian tube, Aga. m. (l.); rushing, Grap., Mag. s.; BEFORE ears, Ant. cr., Arg. n., Asaf., Bry. (l.), Calc. c., Carbn. s., Carb. b., Chel. (l.), Cle., Rhodo.; evening, Caus. (r.);

during mania, (Ars.); during internal coldness, Amyg.; BEGINNING DEEP, then becoming higher, Berb.; DISTANT, All. c. (< r.), Arg. n., Coca, Croc., Led., *Spig.;* intermittent, Mil.; DULL, Pso. (l.), Spo.; FINE, Pul.; FLUTTERING, with hard hearing, Bell., Mag. c., Rhe.; as from high-sounding GLASSES, < evening, Merc. sol.; rattling on a waiter when it is shaken, Zn.; LOUD, driving out of bed, Sil.; like MUSIC, Ail., Pho.; as if ear were SHUT up, then opening with a clap, Mar.; SHRILL, stupefying, sudden, Na. m. (r.); SUDDEN, and ceasing suddenly, Ast. r.

RIVER running, Cact.

ROARING, Acon., Acon. c., Aconin., Aga. m. (l.), (r., then l.), *Agn.*, *Alco.*, Am. car., Ant. cr., Ant. t., Apom., Arn., *Ars.*, Atro., Bar. c., *Bell.*, Benz n., Berb., Bism., Bor., Bov. (l.), Bro. (< l.), *Bry.* (l.), Cai., Cala., Calc. cau., *Calc. c.*, Calth., Camph., Canch., Cannab. s., Carb. ac., Carbn. h., Carbn. o, *Carb. v.*, *Carl.*, Castor. (r.), Caus. (r.), Cham., Chel., *Chin*, Chin. s., Chlf., Cic. v., Cinch, Cit. v., Cle., Coca, Cocc., Coc c (l.), Coff., Coff. t., Colch., Coloc., Con., Con. (r.), Conin., Croc., Crot. t., Cup. ac., *Cyc.*, Daph., Dig., Dir. (r., then l.), Dul., Euo., Fer. i., Gas., Grap. (l.), Guare., Hell., Hell. v., Hep., Hcy. ac., Hyos., Ib., Ig., Ill., Indg., Iod., Jat., K. bi., *K. ca.*, K. clo., K. iod., Kis., Lach., Lact., Lau., Led., Lepi., Lip., Lol., *Lyc.*, Mag. c., Mag. c. (r.), Mag. m., Manc., Mela. (l.), Merc. (r.), Merc. c. (r.), *Merc. sol.*, Mor., Mur. ac. (r.), Na. c., Na. m. (l.), Nicc, *Nit. ac.*, *Nx. v.*, Ol. an., Olnd., *Op.*, Paeo., Pau. p., Pb., *Petrol.*, PHO. AC., Pho., Pimp., Pin. s., *Plat.*, Ple., Pso., Pte., *Pul.*, Rhe. (r.), Rhodo., Rhu. t., *Sal. ac.*, *Sec. c.*, Seneg., Sep., *Sil.*, Spig., Spigg., Spo., *Stap.*, Stram., Stro. (r.), Stry., Sul. ac., SUL., Tab., Tana., Tep., Thea., Ther., Thu. (l.), Til., Wies., Zn. s. Zn.; DAY, Sul.; MORNING, Alum., Calc. c., Merc. sol., Pho. ac., Plat.; in bed, Aur., Na. m.; after rising, Calc. c., *Nx. v.;* on waking, Hype. (< l.); 2 A.M., Pte.; 11 A.M., Mag. c. (r.); AFTERNOON, All. c. (l.), *Ambra*, *Ant. cr.*, Cham.; coming from open air, Thu.; and on rising, Lac. ac. (l.); EVENING, Alum., Caus., Pb., Petrol., Pho. ac., Thu. (l.); in bed, SUL.; after lying down, Plat.; after going to sleep, Calc. c., Sul. ac.; while sitting still, Op.; 3 P.M., Elap., Mag. c. (r.); 8 P.M., Ped.; 9 P.M., Hydrs.; before MIDNIGHT, Euphm.; NIGHT, Am. c. (l.), Cinch., Euphm., *Grap.*, Lepi. (r.); from dreams, Nx. v.; on waking, Con.; from BEATING OF PULSE, Zn. (l.); in BED, Am. car. (r.), Aur., Merc. sol.; on BLOWING nose, Meny. (l.); after EATING, Cinnb., Op.; before MENSES, Bor.; during menses, Bor., Petrol., Verat.; after the NOISE of trumpeting and drumming, Bell.; on RISING, Acon. c., *Pho.;* from a seat, Verat.; while SITTING, Con., Na. m.; at every SOUND, Coloc.; after SPASMS, Alco., *Ars.;* while STRAINING at stool, Lyc.; after STOOPING, Mang.; after SUPPER, Canth.; on WALKING, Colch., Cyc., Fer, Na. m.; from YAWNING, Verat.; (AGG.) towards evening, Pte., Spig.; in bed, Con.; after dinner, Con.; while eating, Con. (l.); in house, Cic. v., Mag. c. (r.); motion, Na. c.; loud noises, Ol. an.; speaking, Na. c.; (AMEL.) sitting up in bed, but returning immediately, Am. car. (r.); swallowing, Rhe. (r.).

—WITH BUZZING and itching, Sep.; DISCHARGE, Bor.; HARD HEARING, Aur., Bor., Croc., Grap., Merc., Seneg., Sep., Thu.; HUMMING, Con. (r. and l.), also in head, Caus., on stooping, Croc.; ITCHING, Sep.; PAIN, Ars.; RELAXED sensation of drum, Rhe.; RINGING, tearing and twitching, from one's own singing, Pul.; RUSHING, Ammc., Am. car., Calc. ostr, Con., Hep., Lyc., Merc., Nit. ac., Pho. ac. Sep., during hot stage, *Ars.*, Bry., Nx. v.; rushing, with shattering pain in head, Plat.; rushing, with fire before eyes, Verat.; STITCHES, Caus., Nit. ac.; STOPPED feeling, Grap., Merc. c., Seneg., Sep.; increased EARWAX, Mur. ac., Sep.; running of wax, Grap.; dulness of SENSORIUM, Arg. n.; HEADACHE, Gel., Hal.; vertigo, Bell., Crot. t., Gran., Hell., Na. c.; heaviness of head, Murx.; cataracts of EYES, Sec. c.; CORYZA, Sep.; swelling of FACE, Cinnb.; pain in ABDOMEN, Bell.; COUGH, All. c., Caus., Con.; FEVER, Lach.

ROARING. FOLLOWED by growling and buzzing, Bell.

— ALTERNATING, with whistling, Mag. c.

— BEATING, Lach.; morning after waking, > shaking finger in ear, Lach.

— BED, driving out of, Mag. c.

— BEFORE, Am. c. (l.), Anac., Bar. c., Card. b., *Caus.*, Dro., Gran., K. n., Merc. sol. (l.); after eating, *Sil.* (l.); morning, after rising, Alum.

— BENUMBING, Bar. c., Olnd.

— BLOOD rushing, like, Petrol., Stan.

— BUBBLING, Aga. m. (l.).

— DEEP-TONED, nights when lying on ear, Sep.

— DISTANT, Pimp., PUL.

— DRAUGHT through a stove, Thu.

— DULL, Arg. n., Til.; every morning, and in evening after lying down, Plat. (r.); in forenoon, Coca; after eating, Op.

— FLUTTERING, of a bird, Plat. (r.); like a partridge, evening on waking, Hydrs.; like a fire in a chimney, a low, dull sound, Bell., Berb., Con., Dul., Lau., Lyc., Mag. c., Olnd., *Petrol.*, Pho., *Plat.*, Rhodo., Spig., Spo., Zn.

— PULSE, synchronous with, *Merc. sol.;* at night, K. br., Sep.; on walking, Acon. c.

— RESOUNDING, Bar. c.

— RIVER, like a, Cact.

— RHYTHMICAL, Coloc. (l.), Sul. ac.

— SEA-SHELL, like, Rum. c. (l.); with hard hearing (after Sul.), Lach.

— SEETHING, Coc. c. (l.).

— STICKING in ear, as if something were, Merc. sol.

— STORM, like a, All. c., Bor., Chel., Coc. c., Con., Led., Mag. c., Pul., Sul., Verat.; storm in a forest, Coc. c.

— SUDDEN, Ast., Bry., Mez. (r.); extending into forehead, > hand over eye, Spig.

— TUBE, as when ear is applied to, Cocc.

— WATER, Ars., Ast., Caus., Cham., Chin. s., Cocc., Con., Mag. c., Mag. s. (l.), Petrol., Pul.; BOILING, Chlf.; RUSHING, *Cham.;* in evening, Petrol.; 7 P.M., Mag. c. (r.); WATERFALL, Ther.; on opening mouth during dinner, Sul. ac.

— WIND, Asar. (l.), Caus., *Chel.*, Con., Croc., *Led.*, *Petrol.*, Verat.; 4.30 P.M., Mag. c. (l.); increasing to a bluster, Acon., Agn., Ambra, Anac., Ant. cr., Arn., *Ars.*, ASAR., *Aur.*, *Bell.*, Bry., Calc. ostr., Cannab. s., Carb. v., Caus., Cic. v., Cocc., Colch., *Con.*, Dro., Fer., *Grap.*, Hep., Ign., K. ca..

Led., Lyc., Mag. c., Mang., Mag. m., Meny., *Mor.*, Murx., Na. c., Nit. ac., *Nx. v.*, Op, Petrol., Pho., Pho. ac., Plat., Pul., Rhe., Rhu. t., Saba., Sec. c., Seneg., Sep., Sil., *Stap.*, *Sul.*, Thu., *Verat.*

ROARING. TWITCHING, at night, Nicc. (r.).

ROLLING, with confusion of head, Zn. cy.; like thunder, Grap.

RUMBLING, Apis, Equ. (l.), Gas.; before ear, Bry. (r.); distant, evening in bed, Sel. (l.); like a wagon, Am. m., Grap., Plat.

RUNNING before ear, in afternoon, Am. car. (l.).

RUSHING, Aga. m., Alco., All. c., Am. car. (r.), Bro. (r.), Caus. (r.), Bap., Bro. (r.), Cap., Dul., Grap., K. ca., K. cy., Lil. t., Mez., Na. c., Pho., Phyt., Rum. c., Sul., followed by Calc. ostr., Tab., Tel., Ther., Verat. v., Vio. o.; MORNING, Dul.; 9.40 P.M., Na. ar. (r.); AFTERNOON, All. c.; after MIDNIGHT, when lying on ear, Am. car. (r.); NIGHT, Ther.; in BED, Na. m.; during COITION, Grap.; on RISING from seat, Verat.; WITH hard hearing, Calc. ostr.; feeling as if something heavy fell on the floor and cracked, after which the noise continues a long time, Saba.; roaring, Ammc., Am. car., Calc. ostr., Con., Hep., Lyc., Merc., Nit. ac., Pho. ac., Sep., during the hot stage, *Ars.*, Bry., *Nx. v.;* roaring with shattering pain in head, Plat.; roaring in head, with better hearing, Galvanism, Lach.; roaring with fire before the eyes, Verat.; whistling, Sep.; dry earwax, Pho. ac.; cough, Dul., Pho.; cold stage, *Ars.*, Pul.; sweat, ARS., *Bell.*, *Calc. ostr.*, *Caus.*, *Grap.*, Hep., Lyc., Nx. v., *Pul.*, Saba., Sep., SUL.; FOLLOWED by amelioration of hard hearing, Lach.; ALTERNATING with ringing, Grap., Mag. s.; as from BLOOD, Stan.; rushing through brain, Con.; DISTANT, Bro.; DISTURBING SLEEP, Euphm, Euphm. (r.); as in FAINTNESS, Mos.; as of FULLING-MACHINE, night, Nx. v.; like a jerking sound of a LOCOMOTIVE, < lying, > rising, Aga. m. (r.); like STEAM escaping, Glo.; night after lying down, Physo.; extending to occiput, Cass.; from a kettle, Glo.; as from a STORM, Bor. (l.); SUDDEN, Na. m.; as from fluttering of a bird, Mos. (r., then l.); as when listening at a tube, Cocc.; as of WATER, *Cocc.*, K. n., Nit. ox.; after 4 P.M., *Pul.;* with increased earwax, Petrol.; falling water, Hydrphb.; waterfall, Ars., Caus.; as of WIND rushing out of ear, Art. ab., Mos., Sul.

RUSTLING, Am. cau., Aur., Bar. c., BELL., Bov., Carb. v., Cham., Chel., *Cocc.*, Con., Dul., K. ca., Mag. c., Mang., Mos. (l.), Na. c., Na. m., Nx. v., Pul., Rhodo., Rhu. v. (r.), Stan., Vio. o.; on moving maxillary joint, Aloe (r.); like a fly, Mos.; like a grasshopper, Stan. (l.).

SCRAPING, scuffling, Cham.

SCREAMING, on blowing nose, Pho. ac., Stan.; shooting through limbs, evening in bed, Grap.

SEETHING, Acon.; like boiling water, Bry., Dig.

SHOOTING, distant, Am. c.; shooting in dreams, *Am. m.*, Hep., Hur., Lach., *Merc.*, Spo.; dream that she was shot, Mag. s.; of a pistol, and a loud sound, in a dream, Cerv.

SHRILL, on blowing nose, Mar., Pho. ac.; stupefying, Ol. an.

SINGING, Acon., Am. m., Arg. n. (r.), Arn. (r.), Asar., Bell., *Calc. c.*, Calc. ostr., Calc. pal., Calc. ph. (< r.), Cannab. i., Carbn. o., Caus., Cer. s. (r.), Chin. s., Coca (l.), Coff. t., Coloc., Cot. (l.), Erig. (r.), Ery. a. (l.), Fer., Fl. ac., Gam., Gas., Glo., Grap., Ir. v., K. bi., *K. ca.*, Lac. ac., Lach., Lachn.,

Lyc., Mar., Mur. ac., Na. m. (l.), Ol. an., Olnd., Op., Pen., Petrol., Phel., Pho., Pso., Pyro. carb., Sang., Sec. c., Sep., Sul. i., Sum., Ter., Vich.; 8 A.M., Phe.; AFTERNOON, when walking in open air, Lachn.; EVENING, Linu., Merc. i. r., Sum. (r.); 5 P.M., Ol. an.; in OPEN AIR, Lach.; CLOSING EYES, Chel.; while LYING, Cannab. i., Pho. ac.; before MENSES, Fer.; during menses, Petrol.; after the bursting NOISE, Saba.; when walking, Lachn.; WITH hard hearing, Pso.; hissing, Gam. (l.); snapping, Lact. ac. (l.); whistling, and looking to see where it came from, Elap.; delirium, Stram.; vertigo, Sang.; spasm of stomach, Atro.; ALTERNATING with throbbing, Caus.; after ringing and tearing, Plat.; BEFORE ears, Lachn., Sep.; as of CRICKETS, Caus., Ced.; DISTURBING SLEEP, Grap.; like a LOCUST, > cold air, afternoon, Rhu. t.; PERIODICAL, *Cannab. i.*; SHRILL, Asar. (r.); like STEAM escaping, after lying down at night, Physo.; like boiling WATER, Lyc., Thu.; like WIND in ear, when walking in open air, Carbn. s. (r.).

SLAMMING, as of a door, Stan.

SNAPPING, Pul. n.; with singing, Lact. ac. (l.); like a harp-string, Sul.

SOUNDING-BOARD, like, when breathing, Bar. c.

SPATTERING, with hard hearing, Plat.; then discharge of moisture, Spig.; then amelioration of hard hearing, Spig.

SPLASHING, like water falling, Ant. cr.; as if filled with water, Grap.

SPUTTERING, Nit. ac., *Plat.*, Sil., *Spig.*

SQUASHING, *Calc. ostr.*, Spig., *Sul.*

SQUEAKING, Eup. pur., Mar.; like young mice, Lyc., Rhu. t.

STEAM escaping. See under RUSHING and SINGING.

STUNNING, Bar. c.

SURRING, Mor.

SWASHING, with cracking, on swallowing, Grap.; as of water, *Sul.*; when moving jaw, Ant. cr.

TALKING, Elap.; after abuse of alcohol in small doses, Ars.; in the corners of the room, Ars.; confused, Benz. ac.; of a strange voice, follows it and tilts against the door, Crot. c.; two persons, in a dream, Stram.

TEA-KETTLE, at a distance, beginning to boil, Aga. m.

THUNDERING, Am. m., Calc. ostr., Caus. (r.), Carbn. o., Chel., Gas., Grap., Lach., Plat., Sil.; at night, Am. m. (r.); while sitting, Am. m. (r.); in dreams, Arn., Ars., Sul.

TICKING, *Chin.*, Dro., Petrol., Ter.; evening, Na. m. (l.); like a distant watch, Cinch.

TICK-TACK, Gad. (r.).

TINKLING (compare with BELLS), Ant. cr., Atro., Calc. ostr., Cle., Croc., Hyos., *Kalm.*, Lau., Led., Mag. c., Mur. ac., Nitr., Osm., Pho. ac., Pul., Spig., Sul. ac., Val.; like little bells, Brom., Bry.; like glasses, Merc.; like a pane of glass breaking, Zn.

TOLLING, as of church-bells, Alum., Am. car., Ars., Calc. ostr., Cle., Con., Hyos., K. ca., K. ca. (r., then l.), Led., Lyc., Mag. c., Mang., Meny., Na. m., Nitr., Phel., Pho., Rhodo., Sars., Sil., *Sul.*, Sul. ac., Val., Zn.; in afternoon, Sars. (l.).

TRUMPETS, Bell., Gas.; and cymbals, Bell.; and kettledrums, Bell.; and drums, then roaring, Bell.

TUMULT, Cocc.; several weeks after wakes with a start during sleep, after midnight, Rhu. t.; a confused sound, after midnight, daily for some weeks, Am. car.

TWANGING, Pul., Saba.

TWITTERING, Cala., Coloc., Euphm., Mur. ac., Nx. v., Pul.; as from a cricket, morning in bed, Pul.; like young mice, Rhu. t.

VOICES. See TALKING.

VIBRATING, Grap., Pho., Tab.; DULL, Carbn. o., Thu., as when an iron rod is struck, Pul.; as of the STRING of an instrument, Cannab. s. (l.).

WABBLING as from water, with over-sensitiveness, Sul.; as if water were running before the ear, in afternoon, Am. car.

WALKING behind him, he hears some one, Crot. c., Bro.; before his bed, at night, Carb. v.

WARBLING, as of birds, Bell., Bry.

WATER, Nit. ac.; BOILING, Bry. (l.), *Cannab. i., Dig.;* before ears, Bry.; out at ears, Sul.; WATERFALL, Cannab. i., Chel., Con. (r.), Na. p., Petros., Sul. ac.; TRICKLING, with sensation of wind in, Thu.

WAVES, Ast.; waving, Fago.; with fretful impatience, Plat.; jumping, as from fleas, Mos.

WHEEL, Cit. s., Hydrs.

WHEEZING, Tarent.

WHIRRING, with confusion of head, Carb. a.

WHISPERING (hallucination), in evening, Rhodi.

WHISTLING, Æth., Alumn., Ambra, Aur., Bell., Bor., Carb. ac., Caus., Chel., Caus. (l.), Elap., Fer., Grap., Hep., Hur., Lyc., Mag. c., Mur. ac., Pul. (r.), Sarr. (r.), Sep., Verat. (r.), Vin., Vin. m.; 9 A.M., Hur. (r.); AFTERNOON, *Ambra;* EVENING when writing, Sep.; when BLOWING NOSE, Carb. a., Hep. (r.), Lyc.; when WALKING, Manc.; WITH ringing, Vin. m.; rushing, Sep.; ALTERNATING with roaring, Mag. c.; DISTANT, Elap.; EXTENDING through every limb, Grap.

WHIZZING, Alum., Arg. n., Berb. (l.), Calc. a., Hur., Lach., Mag. c. (r.), Mim., Naj. (l.), Olnd., Pb., Pho., Sang., Tarent., Thu., Zn. (r.); 11 A.M., Mag. c. (r.); EVENING, Zn.; when writing, Sep.; while LYING in bed, Plat.; when WHISTLING, Ped.; > FOOT-BATH, Tarent.; WITH hard hearing, Mag. c.; rush of blood to head, SANG.; like a PULSATION in head, Spo.

WIND, Carbn. s. (l.), Led., Spig. (l.); 4 P.M., Pul.; 4.30 P.M., Mag. c. (l.); < noise, Plat.; passing rapidly, Spig.; strong, before ear, Ign.; whistling or singing, Vin. m. See also under ROARING.

WINDING OF A WATCH, Ambra.

WINDMILL, in morning, Bry.

INDEX.

www.ingramcontent.com/pod-product-compliance
Lightning Source LLC
Chambersburg PA
CBHW021516210326
41599CB00012B/1272

* 9 7 8 3 3 3 7 2 7 4 6 9 6 *